Dyslipoproteinaemias and Diabetes

Monographs on Atherosclerosis

Vol. 13

Series Editors
Daniel Kritchevsky, Philadelphia, Pa.
O.J. Pollak, Dover, Del.

Basel · München · Paris · London · New York · New Delhi · Singapore · Tokyo · Sydney

European Atherosclerosis Group Meeting, Montreux, March 15–16, 1985

Dyslipoproteinaemias and Diabetes

Volume Editors
R.W. James, Geneva
Daniel Pometta, Geneva

29 figures and 60 tables, 1985

Basel · München · Paris · London · New York · New Delhi · Singapore · Tokyo · Sydney

Monographs on Atherosclerosis

National Library of Medicine, Cataloging in Publication
    European Atherosclerosis Group. Meeting (1985: Montreux, Switzerland)
        Dyslipoproteinaemias and diabetes / European Atherosclerosis Group Meeting, Montreux, March 15–16, 1985; volume editors, R.W. James, Daniel Pometta. – Basel; New York: Karger, 1985.
    (Monographs on atherosclerosis; vol. 13)
    Includes index.
    1. Apolipoproteins – blood – congresses 2. Diabetes Mellitus – physiopathology – congresses 3. Lipoproteins – blood – congresses I. James, R.W. (Richard W.) II. Pometta, Daniel III. Title IV. Series
    W1 MO569T v. 13
    QY 465 E89d 1985
    ISBN 3-8055-4139-2

Drug Dosage
    The authors and the publisher have exerted every effort to ensure that drug selection and dosage set forth in this text are in accord with current recommendations and practice at the time of publication. However, in view of ongoing research, changes in government regulations, and the constant flow of information relating to drug therapy and drug reactions, the reader is urged to check the package insert for each drug for any change in indications and dosage and for added warnings and precautions. This is particularly important when the recommended agent is a new and/or infrequently employed drug.

All rights reserved.
    No part of this publication may be translated into other languages, reproduced or utlized in any form or by any means, electronic or mechanical, including photocopying, recording, microcopying, or by any information storage and retrieval system, without permission in writing from the publisher.

©   Copyright 1985 by S. Karger AG, P.O. Box, CH–4009 Basel (Switzerland)
    Printed in Switzerland by Graphische Anstalt Schüler AG, Biel
    ISBN 3-8055-4139-2

Contents

Introduction	VII
Wilson, Peter W.F. (Framingham, Mass.); *Kannel, William B.* (Boston, Mass.); *Anderson, Keaven M.* (Framingham, Mass.): Lipids, Glucose Intolerance and Vascular Disease; The Framingham Study	1
Kostner, Gerhard M. (Graz); *Schernthaner, Guntram* (Vienna): Apolipoproteins in Diabetes mellitus	12
Breslow, Jan L. (New York, N.Y.): Genetic Determinants of Dyslipidemias	25
Galton, David J. (London): Genetic Polymorphisms in the Analysis of Diabetes mellitus	32
Nikkilä, Esko A. (Helsinki): Very Low Density Lipoprotein Triglyceride Metabolism in Diabetes	44
Flückiger, R. (Basel): Glycosylation of Lipoproteins: Chemistry and Biological Implications	53
Antero Kesäniemi, Y. (Helsinki): Pathophysiology of Low Density Lipoprotein and High Density Lipoprotein Glucosylation	63
Rifkind, Basil M. (Bethesda, Md.): The Lipid Research Clinics Coronary Primary Prevention Trial: Results and Implications	74
Shepherd, James; Packard, Christopher J. (Glasgow): Chlorophenoxyisobutyric Acid Derivatives and Apolipoprotein B Metabolism	85
Epstein, Frederick H. (Zürich): Hyperglycaemia as a Risk Factor for Coronary Heart Disease	92
Hanefeld, M.; Schulze, J.; Fischer, S.; Julius, U.; Schmechel, H.; Haller, H.; and The DIS Group (Dresden): The Diabetes Intervention Study (DIS): A Cooperative Multi-Intervention Trial with Newly Manifested Type II Diabetics: Preliminary Results	98
Weisweiler, P.; Merk, W.; Schwandt, P. (Munich): Analysis of Serum Lipoproteins in Insulin-Dependent (Type I) and Noninsulin-Dependent (Type II) Diabetes mellitus	104
Black, Susan; Brunt, R.V.; Reckless, J.P.D. (Bath): Apolipoprotein E Isoforms in Diabetes	111
Carmena, R.; Ascaso, J.F.; Serrano, S.; Martinez-Valls, J.; Soriano, P. (Valencia): Serum HDL Concentrations in Patients with Type I and Type II Diabetes mellitus	115

Contents

Ferns, G.A.A.; Lanham, J.; Galton, D.J. (London): The Association between Primary Gout and Hypertriglyceridaemia May Be due to Genetic Linkage ... 121

Vessby, B.; Lithell, H. (Uppsala); Lipoproteins and Lipoprotein Lipase Activities in Obese Type 2 Diabetics: Studies of the Relationship between Low Density Lipoproteins and the Skeletal Muscle Lipoprotein Lipase 124

Reckless, J.P.D.; Black, Susan; Brunt, R.V. (Bath): Glycation of Very Low Density Lipoproteins in Diabetes 130

Fievet, C.; Parra, H.; Luyeye, I.; Demarquilly, C.; Fievet, P.; Fruchart, J.C.; Bertrand, M.; Lablanche, J.M. (Lille); *Drouin, P.; Gross, P.* (Toul): Mapping of Lipoprotein Particles with Monoclonal Antibodies in Diabetes and Atherosclerosis 134

Packard, C.J.; Caslake, M.J.; Shepherd, J. (Glasgow): Effects of Fenofibrate on Receptor-Mediated and Receptor-Independent Low Density Lipoprotein Catabolism in Hypertriglyceridaemic Subjects 142

Postiglione, Alfredo; Rubba, Paolo; Cicerano, Umberto; Chierchia, Italia; Mancini, Mario (Naples): Pantethine versus Fenofibrate in the Treatment of Type II Hyperlipoproteinemia 145

Riccardi, G.; Rivellese, A.; Capaldo, B.; Vaccaro, O. (Naples): Influence of Carbohydrate Metabolism on Plasma Lipoprotein Levels 149

Koschinsky, T.; Bünting, C.E.; Rütter, R.; Gries, F.A. (Düsseldorf): Apo B Degradation in Vascular Cells is Regulated by Metabolic Control of Diabetes . 153

Koschinsky, T.; Bünting, C.E.; Rütter, E.; Schütze, R.; Gries, F.A. (Düsseldorf): Diabetic Serum Growth Factor: A New Low Molecular Weight Growth Peptide for Arterial Smooth Muscle Cells of Platelet Origin 159

Labat-Robert, J.; Robert, L. (Créteil): Tissue and Plasma Fibronectin in Diabetes 164

Miskulin, M.; Robert, L.; Robert, A.M. (Créteil): Some Cellular and Molecular Aspects of the Vascular Complications of Diabetes 169

Zöllner, N.; Füessl, H.S.; Goebel, F.D. (München): Pathogenesis and Clinical Relevance of Mönckeberg's Medial Calcinosis 174

Subject Index 177

Introduction

Vascular disease is one of the most frequent complications of diabetes, where many of the major risk factors identified in non-diabetic populations are aggravated. Prominent among these risk factors are lipid disorders, which are also commonly observed in diabetics. This was the theme of a meeting held under the auspices of the European Atherosclerosis Group in Montreux, Switzerland, in March 1985, of which this volume represents the proceedings.

At present a confused picture of dyslipoproteinaemias and diabetes emerges from the limited and somewhat fragmentary nature of research in this area. Thus, any conclusions concerning the contribution of lipid disorders to the occurrence of vascular disease in diabetes is premature. The aim of the meeting was to review the current status of research on, and hopefully contribute to a clearer understanding of lipid disorders and diabetes.

The contributions cover various aspects of the theme 'Dyslipoproteinaemias and diabetes', ranging from the epidemiology of glucose intolerance and vascular disease to genetic considerations arising from the rapidly-developing application of techniques of molecular biology. Also discussed are the effects of treatment of hyperlipidaemia on the aetiology of cardiovascular disease, together with the mechanisms of action of hypolipidaemic drugs. Finally, apoprotein disorders in diabetes and the effect of glycosylation on the metabolism of the three major lipoprotein subclasses are also reviewed.

It is hoped that this volume will be of interest to scientists from different disciplines with interests in atherosclerosis and diabetes.

Lipids, Glucose Intolerance and Vascular Disease: The Framingham Study

Peter W.F. Wilson[a], *William B. Kannel*[b], *Keaven M. Anderson*[a]

[a] Framingham Heart Study, Framingham, Mass., and [b] Boston University School of Medicine, Section on Preventive Medicine, Boston, Mass., USA

Introduction

The atherogenic roles of carbohydrate and lipoprotein metabolism are interwined. Abnormalities of both are commonly seen as precursors to various types of vascular disease [1, 2]. Modest alterations in these atherogenic traits are even seen prior to the diagnosis of diabetes mellitus itself [3]. The present report puts into perspective the relative importance of various measures of glucose abnormality in relation to 30 years of surveillance for the development of cardiovascular sequelae. The importance of blood glucose measured at biennial examinations, apart from any diagnosis of diabetes mellitus or glucose intolerance, is also particularly examined as a risk factor for vascular disease. Prediction of later diabetes mellitus using the standard cardiovascular risk factors is considered. A previous report along this line lacked information on high density lipoprotein (HDL) cholesterol [3]. The availability of this additional information with adequate follow-up has allowed investigation into whether diabetes and cardiovascular disease can both be predicted from a common set of atherogenic risk factors.

Methods

The Framingham cohort population under study consisted of 5,209 men and women age 30–62 years at the onset of the study in 1949. Participants were seen at 2-year intervals and clinical vascular disease end points were diagnosed from exam information, hospital records, medical examiner's reports, and the patient's personal physician notes.

Diagnostic criteria for the various end points are given elsewhere [4]. For some analyses reported herein the various clinical components were grouped. Coronary heart disease (CHD) included angina pectoris, coronary insufficiency, and myocardial infarction. Cardiovascular disease (CVD) included all the CHD groupings, as well as congestive heart failure, intermittent claudication, stroke and transient ischemic attack.

The diagnosis of 'diabetes mellitus' was made when any one of several criteria were met: (a) history of treatment with insulin or an oral hypoglycemic agent; (b) finding a casual blood glucose level greater than 8.33 mmol/l (150 mg/dl) on two visits to the heart study; (c) a positive glucose tolerance test using the Fajans criteria. Individuals with blood sugar abnormalities but without diabetes mellitus were diagnosed as having 'glucose intolerance' if any of the following criteria were met: (a) casual blood glucose of 6.67 mmol/l (120 mg/dl) or greater; (b) glucosuria with a reading of trace or above; (c) an abnormal 100 g oral glucose tolerance test using Fajans-Conn criteria of greater or equal to 8.88 mmol/l (160 mg/dl) at 1 h, greater or equal to 6.11 mmol/l (110 mg/dl) at 2 h, and above basal at 3 h.

Blood glucose was measured by the Somogyi-Nelson method on nonfasting patients at the time of their regular biennial examinations [5]. Blood lipid measurements including HDL cholesterol were done at the twelfth examination using a modification of the Lipid Research Clinic Program procedure. Determinations included very low density lipoprotein (VLDL) cholesterol and low density lipoprotein (LDL) cholesterol, using a protocol including ultracentrifugation [6]. Cholesterol was measured by the Abell-Kendall method in serum at the first eleven exams and from plasma thereafter [7]. Triglyceride was measured by the Kessler-Lederer method at the twelfth examination [8]. HDL cholesterol was measured after precipitation with heparin-manganese [6].

Incidence was calculated for the 30-year rates using a cross-sectional pooling method, with observations made every 2 years. Individuals were reclassified at the start of each biennial interval as to current age, risk factor status, and presence or absence of disease in question. Logistic regression analysis using the Duncan-Walker method was employed for the 30-year vascular disease risk factor relationship calculations [9]. This technique was also employed for the 8-year diabetes incidence which used risk factor data from the twelfth examination as a baseline including the detailed lipoprotein determinations available only after that time. Comparisons involving age-adjusted variables were made using the Student's t test. A general linear model was used to age adjust continuous variables, and a logistic model was employed for dichotomous variables. In each case, age adjustment used the diabetic population as the standard.

Results

The number of events and observations comprising the 30-year follow-up period for new cases of CHD and first occurrence of cerebrovascular disease (stroke or transient ischemic attack) are shown in table I. Each observation interval actually represents 2 years of follow-up as new cases were ascertained biennially. A smaller number of

Table I. Vascular events and observations – Framingham Heart Study, 30-year follow-up

Age Group	Coronary heart disease		Stroke or TIA	
	men	women	men	women
35–64				
Events	515	304	100	85
Observations	20,160	26,996	21,462	27,596
65–94				
Events	240	266	127	160
Observations	4,936	9,284	5,865	9,202

Table II. CHD incidence rates – Framingham Heart Study, 30-year follow-up

Age group	Risk factor	Glycosuria		Glucose intolerance		Diabetes mellitus	
		women	men	women	men	women	men
35–64	absent	6	12	5	12	5	12
35–64	present	21**	26**	14***	19*	19***	20*
65–94	absent	15	23	15	23	15	24
65–94	present	55***	65***	28***	31	32***	34

All incidence rates age-adjusted per 1,000 per year.
Key: significance tests for comparisons within age group for presence or absence of risk factor after multivariate adjustment.

observations were available for determination of rates in those over 65 years, but in this older group events were much more common.

The occurrence of new cardiovascular events grouped according to whether an individual had glycosuria, glucose intolerance, or diabetes mellitus is shown in table II for CHD and in table III for first occurrence of cerebrovascular disease. The format is similar in each circumstance, showing sex-specific age-adjusted disease rates per 1,000 per year for a younger group (35–64 years) and an older group (65–94 years). Significance tests evaluate the effect of a glucose abnormality compared to normals of the same sex within that age group. All statistical tests were made after adjustment for age, systolic blood

Table III. Stroke or TIA incidence rates – Framingham Heart Study, 30-year follow-up

Age group	Risk factor	Glycosuria		Glucose intolerance		Diabetes mellitus	
		women	men	women	men	women	men
35–64	absent	1	2	1	2	1	2
35–64	present	7*	4	3	5	3*	6*
65–94	absent	8	11	7	10	7	10
65–94	present	19***	5	17**	15	19***	17*

All incidence rates age-adjusted per 1,000 per year.
Key: significance tests for comparisons within age group for presence or absence of risk factor after multivariate adjustment.

pressure, cigarettes smoked per day, serum cholesterol, and the presence or absence of left ventricular hypertrophy by electrocardiogram.

As seen in table II, all measures of a glucose abnormality show significantly higher CHD rates in both younger and older women. When incidence rates are compared, the relative risks for coronary disease in Framingham women with a glucose abnormality averages about three in the younger group, and about two in older women. The pattern for men with glucose abnormalities is not as uniform. The most consistent finding is seen for glycosuria, where the highest rates of coronary disease and greatest risk ratios are observed, and a significant difference from normal patients is seen for both younger and older men.

Figures 1 and 2 show the relationship between casual blood glucose and later CHD in 35–64 and 65–94 year groups respectively. Each bar represents a rate which has been adjusted for age. A significant independent effect for casual serum glucose is seen for women in both age groups. A significant independent effects for casual serum glucose in men applies only to the older group. The glucose levels are modest; the top level represents values above 7.3 mmol/l (130 mg/dl).

Among diabetics, first occurrence of cerebrovascular disease, indicated in table III, occurred at lower rates than for coronary disease; the relative infrequency of such disease in persons under 65 years contrasts to the situation for coronary disease. Nevertheless, significant excesses

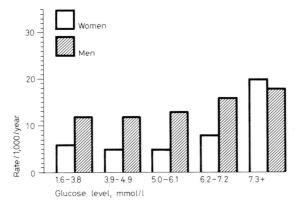

Fig. 1. Age-adjusted CHD incidence per 1,000 per year for men and women 35–64 years according to casual blood glucose. Trend is significant for women at p<0.001.

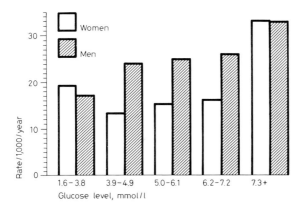

Fig. 2. Age-adjusted CHD incidence per 1,000 per year for men and women 65–94 years according to casual blood glucose. Trends are significant for women at p<0.001, and for men at p<0.01.

of cerebrovascular disease rates were seen for diabetic men and women 35–64 years compared to the nondiabetic group in the same age range. A similar excess risk was seen for subjects over 65 years of age with diabetes. Older women with an abnormality of glucose metabolism were particularly at increased risk for cerebrovascular disease.

Comparisons of age-adjusted mean levels of lipids and other cardiovascular risk characteristics measured at the eleventh examina-

Table IV. Mean levels of lipids and other characteristics – Framingham Heart Study

Variable	Men		Women	
	normal patients	diabetic patients	normal patients	diabetic patients
Total cholesterol, mg/dl	218	214	243	241
Triglyceride, mg/dl	133	161**	118	150***
VLDL cholesterol, mg/dl	31	37**	29	35***
LDL cholesterol, mg/dl	143	135*	157	154
HDL cholesterol, mg/dl	45.7	42.4*	57.0	51.4***
Total Chol/HDL-C ratio	5.20	5.39	4.56	5.10***
Cigarette Use, %	18	17	17	18
Body mass index, kg/m^2	26.5	27.7**	26.3	28.3***
Systolic pressure, mm Hg	141	150***	146	157***
Number of patients	1,074	130	1,449	135

Note: significance tests for age-adjusted sex-specific comparisons between diabetic and normal patients.

tion are shown in table IV. Consistent differences between diabetic and nondiabetic patients were seen for triglyceride, VLDL cholesterol, HDL cholesterol, the total cholesterol/HDL cholesterol ratio, body mass index, and systolic blood pressure.

From the standpoint of abnormal lipid values, the frequency of lipid extremes using the Lipid Research Clinic Program's criteria for high and low levels of various lipids are shown in table V [10]. The definitions used for a high level of a factor were the sex- and age-specific 90th percentile; for a low level the sex- and age-specific 10th percentile levels were used. All cells in table V would be expected to have entries close to an expected 10%. Such results were not obtained, even in the nondiabetic normal men and women in Framingham, all of whom were over 50 years at the time of the determinations. Comparing diabetic patients to normal individuals of the same sex, a significantly greater proportion with high triglyceride and low HDL cholesterol were seen in the diabetics of both sexes.

Tables VI and VII display mean values of characteristics in men and women which were investigated as possible predictors of diabetes mellitus. This analysis uses the twelfth exam as a baseline which in-

Table V. Frequency of lipid extremes, Lipid Research Clinic Program criteria – Framingham Heart Study, twelfth examination

Lipid abnormality	Men		Women	
	normal %	diabetic %	normal %	diabetic %
High total cholesterol	14	13	21	24
High triglyceride	9	19**	8	17***
High VLDL cholesterol	26	34*	31	38
High LDL cholesterol	11	9	16	15
Low HDL cholesterol	12	21**	10	25***
High HDL cholesterol	7	3	4	5
Number of subjects	1,074	130	1,449	135

Note: significance tests for sex-specific comparisons between diabetic and normal patients. All cells 10% in LRC data.

Table VI. Factors related to the incidence of diabetes mellitus (in males) using logistic model – Framingham Heart Study, 8-year follow-up

Variable	Means		Univ. signif.	Multiv. signif.
	cases	noncases		
Diuretic use, %	49	32	***	
History of CVD, %	29	16	***	
Total cholesterol/HDL-C	6.07	5.14	***	**
Cigarette use, %	23	25		
Body mass index, kg/m^2	28.3	26.5	***	**
Systolic pressure, mm Hg	146	139	**	
Age, years	61	61		
Number of patients	55	906		

cluded lipoprotein quantification and allowed 8 years of follow-up. Diuretic use at the time of the baseline exam or during follow-up was also included. Similarly, CVD present at the baseline was considered as a potential precursor. Although several variables are significantly associated with the occurrence of diabetes in the univariate situation, only the total cholesterol/HDL cholesterol ratio and body mass index

Table VII. Factors related to the incidence of Diabetes mellitus (in females) using logistic model – Framingham Heart Study, 8-year follow-up

Variable	Means		Univ. signif.	Multiv. signif.
	cases	noncases		
Diuretic use, %	84	45	***	**
History of CVD, %	13	11		
Total cholesterol/HDL-C	5.43	4.48	***	**
Cigarette use, %	13	22		
Body mass index, kg/m^2	30.1	26.1	***	***
Systolic pressure, mm Hg	151	142	**	
Age, years	62	62		
Number of patients	38	1,282		

remain as important predictors of diabetes mellitus in men and women in the multivariate model. For women alone diuretic use was also found to be an independent predictive factor.

Discussion

The data reported confirm once again a relationship between various indicators of impaired glucose tolerance and the development of CHD and stroke [11]. The definitions of 'glucose intolerance' or 'diabetes' are not precisely those accepted by the American Diabetes Association, but do reflect phenomena associated with grossly disordered glucose tolerance. The Framingham methods may be criticized for lack of specificity when measured against rigorous criteria for 'diabetes'. However, these comparatively low threshold values do identify persons with a two- to threefold increased risk of major cardiovascular events. This suggests that in relation to atherogenesis, a modest derangement in glucose metabolism may be important. This is borne out by the data relating casual blood glucose to cardiovascular disease. Figures 1 and 2 display this relationship, and indicate that blood glucose values usually regarded as 'nondiabetic', or within the normal range, may be associated with an increased risk of vascular disease. This finding of a graded effect without a true critical value is reminiscent of the suggestion with 'hypertension' and 'hyperlipidemia'. The results are

Table VIII. Multivariate risk quintiles, coronary heart disease incidence – Framingham Heart Study, 8-year follow-up

Risk quintile	Men		Women	
	normal patients %	diabetic patients %	normal patients %	diabetic patients %
1	9.2	0.0	10.1	13.3
2	18.5	20.0	13.0	13.4
3	21.3	15.0	23.2	6.7
4	22.4	35.0	21.7	33.4
5	27.6	30.0	31.9	33.3
Total cases	152	20	138	15
Number at risk	981	121	1,349	111

Note: after adjustment for systolic pressure, smoking, total cholesterol/HDL cholesterol ratio, age, and body mass index in the multivariate regression model to calculate risk quintiles.

at odds with claims that it is 'diabetes' but not 'hyperglycemia' which predisposes to cardiovascular disease [2, 12, 13]. Recent data from the Bedford Study are also supportive of glucose intolerance as a risk factor for cardiovascular mortality [14], and support this view.

It is of interest that many of the risk factors for CVD are also risk factors for diabetes. A noteworthy fact is that many of the lipid derangements found in uncontrolled diabetes also predict diabetes. These shared risk factors apparently predict CVD as well in the diabetic as in the nondiabetic, as seen in table VIII which compares risk factor quintiles for diabetics and nondiabetics. At higher levels of risk, the diabetic has a greater chance than the nondiabetic of developing CHD.

The nature of the unique contribution of diabetes to CVD is not well understood. Associated lipid and other risk factors undoubtedly account for a large part of the excess risk of diabetes, but not all of it. Lipid aberrations common to diabetes are major controllable contributors. Some data suggest that aberrations in the clotting mechanism in diabetics may account for some of the unique effects [15].

Insidious deterioration of lipid and carbohydrate metabolism often occur with advancing age. These changes tend to occur in parallel and jointly contribute to the atherosclerotic process. Not only does diabetes contribute to CVD, but CVD also predisposes to diabetes, strongly suggesting a common substrate. Peripheral vascular disease, the most strongly related vascular sequela of diabetes, has been shown to also be a risk factor for diabetes in the Israeli Heart Study as well as the Framingham Study [3, 16]. The situation in such prediction can be complex, though. For instance, interim diuretic use was considered as a potential predictor of diabetes mellitus in Framingham patients, and a significant effect was seen for women. This finding is not unexpected in view of the direct insulin inhibitory effect of diuretics on the pancreas.

The commonality of risk factors, such as the total cholesterol to HDL cholesterol ratio and body mass index for diabetes and CVD, suggest that preventive measures directed against these risk factors should be effective in controlling these two diseases. Control of diabetes for avoidance of CVD evidently should include normalizing the blood lipids and other cardiovascular risk factors in addition to normalization of the blood glucose.

References

1 Kannel, W.B.; McGee, D.L.: Diabetes and cardiovascular risk factors: The Framingham Study. Circulation *59:* 8–13 (1979).
2 Fuller, J.H.; Shipley, M.J.; Rose, G.; Jarrett, R.J.; Keen, H.: Coronary-heart-disease risk and impaired glucose tolerance: The Whitehall Study. Lancet *i:* 1373–1376 (1980).
3 Wilson, P.W.; McGee, D.L.; Kannel, W.B.: Obesity, very low density lipoproteins, and glucose intolerance over fourteen years. The Framingham Study. Am. J. Epidem. *114:* 697–704 (1981).
4 Shurtleff, D.: Some characteristics related to the incidence of cardiovascular disease and death. Framingham Study, 18-year follow-up. DHEW publ. No. (NIH) 74–599 (1974).
5 Nelson, N.: A photometric adaptation of the Somogyi method for the determination of glucose. J. biol. Chem. *153:* 375–380 (1944).
6 Manual of Laboratory Operation, Lipid Research Clinics Program, vol. 1. DHEW publ. No. (NIH) 75–628 (May 1974).
7 Abell, L.L.; Levy, B.B.; Brodie, B.B.; Kendall, F.E.: A simplified method for the estimation of total cholesterol in serum and demonstration of its specificity. J. biol. Chem. *195:* 357–366 (1952).
8 Kessler, G.; Lederer, H.: Fluorometric measurements of triglycerides; in Skeggs, Automation in analytical chemistry. Technicon Symposium, pp. 341–344 (Mediad Inc., New York 1965).

9   Walker, S.H.; Duncan, D.B.: Estimation of the probability of an event as a function of several independent variables. Biometrika *54:* 167–179 (1967).
10 Rifkind, B.M.; Segal, P.: Lipid Research Clinics Program reference values for hyperlipidemia and hypolipidemia. J. Am. med. Ass. *250:* 1869–1872 (1983).
11 Gordon, T.; Castelli, W.P.; Hjortland, M.C.; Kannel, W.B.; Dawber, T.R.; Diabetes, blood lipids, and the role of obesity in coronary heart disease risk for women: The Framingham Study. Ann. intern. Med. *87:* 393–397 (1977).
12 Ducimetiere, P.; Eschwege, E.; Richard, J., et al.: Relationship of glucose tolerance to prevalence of ECG abnormalities and to annual mortality from cardiovascular disease: results of the Paris prospective study. J. chron. Dis. *32:* 759–766 (1979).
13 Stamler, R.; Stamler, J.; Lindberg, H.A., et al.: Asymptomatic hyperglycemia and coronary heart disease in middle-aged men in two employed populations in Chicago. J. chron. Dis. *32:* 805–815 (1979).
14 Jarrett, R.J.; McCartney, P.; Keen, H.: The Bedford Survey: ten-year mortality rates in newly diagnosed diabetics, borderline diabetics and normoglycaemic control and risk indices for coronary heart disease in borderline diabetics. Diabetologia *22:* 79–84 (1982).
15 Mustard, J.F.; Packham, M.A.: Platelets and diabetes mellitus. New Engl. J. Med. *297:* 1345–1347 (1977).
16 Medalie, J.H.; Papier, C.M.; Goldbourt, U.; Herman, J.B.: Major factors in the development of diabetes mellitus in 10,000 men. Archs intern. Med. *135:* 811–817 (1975).

Peter W.F. Wilson, MD, Framingham Heart Study, 118 Lincoln Street, Framingham, MA 01701 (USA)

Apolipoproteins in Diabetes mellitus[1]

Gerhard M. Kostner[a], Guntram Schernthaner[b]

[a] Institute of Medical Biochemistry, University of Graz, and
[b] Second Department of Medicine, University Hospital, Vienna, Austria

Introduction

It is a widely accepted fact that individuals suffering from diabetes mellitus (DM) are at an increased risk for atherosclerosis and myocardial infarction. In fact it seems that diabetes is one of the 5 most cited risk factors. Among the different forms of vascular diseases, peripheral atherosclerosis certainly is the most common secondary complication in that context.

Numerous theories have been put forward which possibly may explain the connection of diabetes with atherosclerosis, among them basement membrane thickening followed by microangiopathy, disturbances in the prostanoid metabolism, platelet dysfunctions and many others. Many of these abnormalities have generally been found in hyperlipoproteinemia and atherosclerosis not connected to diabetes.

In this report we will restrict ourselves predominantly to abnormalities in DM linked to the lipid metabolism with major emphasis to apolipoprotein alterations.

The Major Apolipoprotein in Human Plasma

Table I summarizes the most important Apo-Lp. They are divided according to the lipoprotein family concept [1] into classes A through

---

[1] Parts of this study were supported by grants from the Austrian Research Foundation, grant No. P5144.

Table 1. Apolipoproteins of human plasma

Apo-Lp	Density class	Function
AI	HDL	activation of LCAT, transport of polar lipids
AII	HDL	activation of hepatic lipase
B	VLDL, LDL, Lp(a)	mediates lipid absorption, receptor-mediated catabolism of LDL
CII	CYM, VLDL	activates LPL
CIII	CYM, VLDL	involved in TG clearance
E	VLDL, IDL	binds to specific E receptor, cause CYM remnant clearance
H	CYM, d 1.21 bott.	involved in the TG lipolysis
Lp(a)	HDL_1	unknown

H. For a more detailed review see *Kostner* [2]. ApoAI, the major HDL protein, is responsible for the transport of polar lipids and also activates the enzyme lecithin:cholesterol acyltransferase (LCAT). ApoB is responsible for the lipid absorption and secretion into the circulation and also for the catabolism of cholesteryl ester-rich lipoproteins, e.g. LDL, IDL and Lp(a). The C proteins are responsible for triglyceride (TG) hydrolysis by lipoprotein lipase (LPL). ApoE mediates the catabolism of VLDL and chylomicron remnants through the specific E receptor and ApoH may be involved in the lipolytic system. Lp(a), one of the most atherogenic lipoproteins [3] has not been characterized with respect to function and physiologic significance. There seems to be little doubt that all ApoB-100 containing lipoproteins, VLDL, IDL, LDL and Lp(a), are of atherogenic nature and that increased plasma concentrations are linked to premature vascular diseases. High concentrations of Apo-A containing Lp on the other hand either protect from atherosclerosis or at least signalize a metabolic situation which may be protective. ApoC is necessary for the catabolism of TG-rich Lp but information available so far suggests that disturbances in ApoC plasma concentrations (excepting genetic deficiencies) do not signalize any atherogenic situation. There are, however, indications that in hypertriglyceridemia, there is a shift in ApoC distribution from HDL towards Lp of $d < 1.063$ due to a reduction in VLDL catabolism. Similar findings were reported for ApoE [4]: This Apo-Lp might either be increased in concentration (e.g. type III HLP) or shifted towards VLDL and IDL, indicating an increased atherosclerosis risk.

Table II. Literature survey of apolipoproteins in DM

Type of DM	ApoAI	ApoAII	ApoB	ApoCII	ApoCIII	ApoE	Ref.
I	–	–	↑	↔	↔	–	5
I	↑	↔	–	–	–	–	6
I	↑	↔	↔	–	–	–	7
I	↓	–	↑	–	–	–	8
I+II	↓	↓	–	↑	↑	–	9
II	–	–	–	↑	↑	↔	10
II	–	–	–	↔	↔	–	11

↑ = increased ; ↓ = reduced ; ↔ = not changed; – = not investigated.

Apolipoproteins and Diabetes

In the following report we will try to answer four basic questions: (1) Are there some quantitative differences in plasma Apo-Lp concentrations? (2) Do some qualitative differences in Apo-Lp exist? (3) Are there some differences between individuals suffering from type I or from type II diabetes? (4) Can the observed disturbances of plasma Apo-Lp explain the increased atherosclerosis risk and what are the therapeutic consequences?

Plasma Apo-Lp Concentrations in DM

Apo-Lp have been measured in diabetic individuals since approximately 10 years and this is documented in a great number of reports. Many of these reports are contradictory in the sense that all kinds of alterations have been found. Some authors report on increases, others of reductions and even others on no changes in Apo-Lp of the A, B or C class. This may have methodological reasons on the one hand, but certainly also reflects the metabolic control of the patients. Table II gives some examples of these divergent findings. Because of the inability to evaluate these and all the other studies on that topic in detail, we will restrict ourselves in the following primarily to our own research in this area [12].

We have studied more than 80 diabetic individuals and divided them according to sex and type of the disease into type I, i.e. juvenile insulin-dependent DM (IDD), and type II, i.e. maturity-onset, noninsulin-dependent diabetics (NIDD). For each group, a control group

Table III. Clinical and biochemical characteristics of diabetic patients and of controls

	Diabetics		Controls	
	males	females	males	females
Insulin-dependent (type I) diabetics				
Number	24	20	20	18
Age, years (mean)	35.7	27.4	31.1	28.7
Duration of DM, years	14.8	11.6	–	–
IBW, %	94.1	96.7	96.2	89.5
FBG, mg/dl (mean)	224	261	83	77
HbA_1 %	10.0	10.6	7.3	7.2
Noninsulin-dependent (type II) diabetics				
Number	18	20	18	18
Age, years	61	64	62	57
Duration of DM, years	7	5	–	–
IBW, %	109	113	106	107
FBG, mg/dl (mean)	175	193	88	86
HbA_1, %	9.5	9.8	7.2	7.2

matched for age and socioeconomic status was investigated. Some of the characteristics of patients and controls are listed in table III. Alcohol, cigarette smoking and the degree of physical exercise were similar in patients and controls. Women using oral contraceptives and patients with retinopathy or nephropathy (plasma creatinine > 1.2 mg/dl) were excluded.

Clinical Chemical Analyses

Total cholesterol (TC) and free cholesterol (FC) and TG were measured enzymatically with Boehringer and Merck enzyme kits. HDL-C was measured in the supernatant from phosphotungstate/Mg^{++} precipitation and LDL-C was calculated by use of the Friedewald formula. VLDL-TG were measured in the supernatants after ultracentrifugation with an airfuge (Beckmann). Apolipoproteins AI, AII, B and Lp(a) were quantified immunochemically using own antibodies or commercial preparations from Immuno AG, Vienna, Austria, as described [13]. LCAT was determined by measuring FC immediately after blood drawing and following 30 min incubation at 37 °C [14].

Table IV. Plasma lipid and lipoprotein concentrations in type I and type II diabetic patients and in controls (values are means and given in mg/dl)

	Type I	Controls	Type II	Controls
Males				
TC	233	214	230	241
LDL-C	156	138	144	161
HDL-C	39*	45*	39*	48*
TG	145	141	229	193
VLDL-TG	87	96	171	133
FC	56	51	65	65
LCAT	83	83	96	94
Females				
TC	214	189	237	220
LDL-C	135***	108***	166	144
HDL-C	49***	62***	40***	51**
TG	120	84	197**	123**
VLDL-TG	62	52	132**	77**
FC	50	43	62	58
LCAT	83	82	102	88

* = $p<0.05$; ** = $p<0.01$; *** = $p<0.001$.

Evaluation of the Metabolic Control of Diabetic Patients

The state of metabolic control was evaluated by measuring fasting blood glucose (FBG), 24-hour glucosuria and HbA_1 concentrations (Clinitest). None of the patients showed ketonuria at the time of examination. The patients were divided into three groups (A, B, C) according to the degree of diabetic control. Type I diabetics were considered in good metabolic control (group A) when their urinary glucose excretion was less than 5% of total dietary available glucose. Fair control (group B) was considered between 5 and 10% and poor control > 10% urinary glucose excretion. Most of the type I diabetics of group I had HbA_1 values of < 9%, group B between 9 and 12% and group C > 12%. Type II diabetics were classified as A with < 130 mg/dl FBG, as B with 130–200 mg/dl FBG and as C with > 200 mg/dl FBG. Table IV lists the blood lipid and Lp lipid values of patients and controls in addition to the significance of possible differences. HDL-C was reduced in type I and type II male and female diabetics. We also observed an increase in total TG and VLDL-TG in female type II

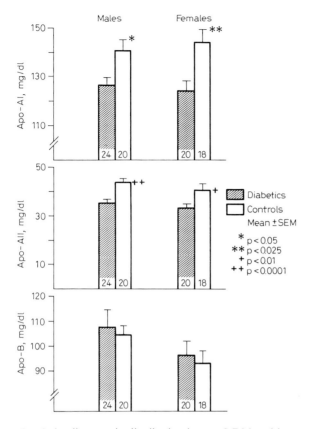

Fig. 1. Apolipoprotein distribution in type I DM and in controls.

diabetics and an increase in LDL-C in female IDD. All other lipids, lipoproteins and also LCAT were not significantly different between patients and controls.

Figures 1 and 2 show the distribution of Apo-Lp AI, AII and B: AI and AII were significantly reduced and ApoB in most cases increased in DM patients. The atherosclerosis index AI/B or AII/B was significantly lower in all types of DM as compared to control individuals. Figure 3 shows the values of Lp(a). In this case all diabetics were combined and compared to a normal control collective. Although there was a tendency of increased Lp(a) in diabetics, this increase did not reach statistical significance.

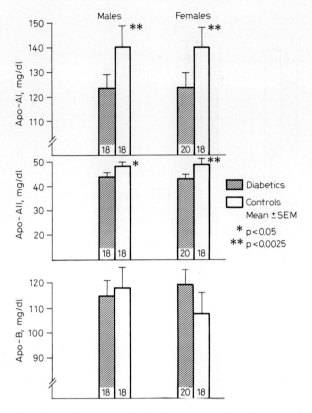

Fig. 2. Apolipoprotein distribution in type II and in controls.

We have also correlated all parameters with the degree of metabolic control in type I and type II DM. Only ApoAII was significantly negatively correlated with urinary glucose excretion. In type I diabetics total TG and log VLDL-TG were significantly correlated with HbA_1. When the daily insulin dosage was correlated with the different parameters, we found a positive correlation with ApoAI, AII and HDL-C. From these studies we conclude that there are mainly quantitative but little detectable qualitative changes in the lipoprotein and Apo-Lp distribution in DM. The most striking are the increase in VLDL and the decrease in HDL. ApoAI and AII are significantly reduced and ApoB, mostly VLDL-B, were increased. No differences have been found between type I and II DM.

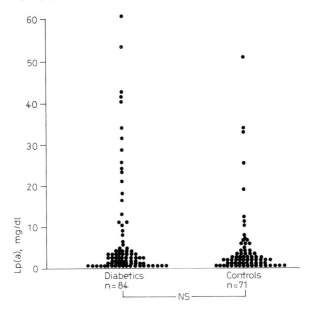

Fig. 3. Plasma Lp(a) values in diabetics and in controls. Type I and type II diabetics were combined.

In another study the effect of normoglycemia on plasma lipoproteins and LCAT activity was studied. 9 nonketotic IDD were controlled for 2 weeks by intravenous insulin pump treatment (IVIPT). The mean age of patients was 27.6 years and the mean duration of the disease was 9.7 years. Before IVIPT, HbA_1 was 13.2% and mean blood glucose levels 230 mg/dl. Patients had been advised to remain on their usual diet consisting of 45% CH, 18% PR and 37% FA. After 2 weeks, IVIPT HbA_1 was significantly reduced to 9.7% and mean blood glucose to 102 mg/dl. The lipid and Lp changes occurring during pump treatment are shown in table V. Total plasma TG and VLDL-TG decreased and ApoAI and ApoAII increased significantly during IVIPT. All other measured lipids and Apo-Lp remained constant. LCAT rose slightly but not significantly.

From these results we conclude that those Lp and Apo-Lp which have been shown to be quantitatively altered in DM can be normalized within a short period of treatment with intravenous insulin and optimal metabolic control. The atherosclerosis index AI/B which was abnormally low before treatment also normalized within 2 weeks.

Table V. Lipid and lipoprotein values of 9 individuals suffering from type I diabetes, treated for 2 weeks with an insulin pump (values are means and given in mg/dl)

Parameter	Pre-pump	Pump therapy	p
Glucose	232	102	<0.001
HbA_1, %	13.2	9.7	<0.001
TG	162	95	<0.05
TC	170	188	n.s.
LDL-C	96	113	n.s.
HDL-C	41	49	<0.05
FC	46	48	n.s.
LCAT	96	113	n.s.
ApoAI	113	149	<0.01
ApoAII	39	47	<0.01
ApoB	80	79	n.s.
ApoE	3.7	4.0	n.s.
Lp(a)	16	16	n.s.
ApoAI/ApoB	1.59	1.99	<0.01

In a final study the action of lipid-lowering drugs on the disturbed Lp pattern in DM was investigated [15]. 16 hyperlipemic, diet-resistant NIDD were treated for 8 weeks with either placebo or with daily 3×200 mg of bezafibrate (BF) (Boehringer, Mannheim, FRG). Diet and dosage of oral antidiabetic drugs (glibenclamide) were kept constant during the study. Lipids, Lp and LCAT were measured after the placebo and after the BF period. Body weight, liver and kidney function tests remained unchanged during the study. The changes in Lp are listed in Table VI. TG, VLDL-TG and TC values were significantly reduced and HDL-C increased by BF. BF also caused an increase in ApoA and a decrease in ApoB and LCAT activity. LDL-C/HDL-C was significantly reduced and AI/B significantly increased by BF.

This study indicates that NIDD individuals which seem to be rather resistant to diet and oral antidiabetic therapy with respect to hyperlipoproteinemia respond very well to lipid-lowering drugs. In many cases, the plasma Lp pattern could be normalized within 8 weeks of treatment.

Table VI. Effect of bezafibrate therapy on plasma lipid and lipoproteins in DM: 16 hyperlipidemic type II diabetics were treated for 2 months with placebo and for 2 months with 2 × 300 mg/day of bezafibrate and lipoproteins were measured (values are means and given in mg/dl)

Parameter	Placebo	Bezafibrate	p
TG	256	192	<0.01
VLDL-TG	183	128	<0.05
TC	260	235	<0.05
LDL-C	166	149	n.s.
HDL-C	40	45	<0.01
ApoAI	121	127	n.s.
ApoAII	34	44	<0.001
ApoB	100	92	n.s.
LCAT	141	124	n.s.
LDL-C/HDL-C	4.1	3.4	<0.01
ApoAI/ApoB	1.2	1.4	<0.01

General Considerations

Results from our studies and partially also from previous investigations have shown that DM is connected with abnormal plasma lipid and Lp concentrations. The most striking abnormalities are the rise in total and VLDL-TG and the decrease in HDL-C and ApoA. From metabolic studies reviewed by *Nestel* [16] it is known that the hypertriglyceridemia is mainly caused by an increase of the VLDL synthesis rate. Thus, overproduction of TG-rich Lp account mainly for the hyperlipidemia. On the other hand, poor metabolic control of DM has frequently been found to be connected with reduced LPL in adipose tissue. This LPL activity, however, normalizes upon optimal metabolic control.

The well-documented inverse correlation between VLDL and HDL found in the nondiabetic population has also been verified in diabetics. The same mechanisms acting in the former group, namely the reduced LPL activity seems to be responsible for this reduction in HDL. Here, not only HDL-C but even more ApoAI and ApoAII were significantly reduced. ApoB in VLDL are increased and in LDL mostly decreased. This seems to be also caused by a reduced VLDL flux (to

LDL) by reduced LPL activities. Because of the increase in VLDL mass, the other Apo-VLDL proteins of the C and E family are also increased but only in the TG-rich fraction. A concomitant reduction in HDL-ApoC and HDL-ApoE is a common feature in poorly controlled diabetics.

Concerning qualitative differences of Apo-Lp in DM, the increased glucosylation of ApoB will only briefly be mentioned, since this is the topic of another report of this symposium. Since nonenzymatic protein glucosylation affects the NH_2 groups of Lys and this latter amino acid plays a key role in the receptor-specific LDL catabolism, poor metabolic control of DM may also impair through this route the LDL catabolism. In animal experiments it has been shown [17] that the glucosylation of ApoB by 5% reduces its catabolism up to 25%. Similar considerations may also apply for ApoE.

In summary, we have shown in this investigation that DM is accompanied by hyperlipoproteinemia primarily of TG-rich lipoproteins. This is caused by an increased VLDL synthesis rate and by a reduced catabolism by LPL. Similar to nondiabetics, the increase of VLDL causes a reduction of HDL. The main Lp density classes, VLDL, LDL and HDL, however, seem to be normally composed except for possible alterations by nonenzymatic glucosylation of ApoB. Thus, an increase in VLDL is accompanied by a rise in VLDL-ApoB, -ApoC and -ApoE and the reduction in HDL reflects also the reduction in ApoAI and ApoAII. Lp(a) was not significantly increased in DM, but further studies will be necessary to clearly confirm these preliminary observations. From our study it also became apparent that the lipoprotein abnormalities correlate with the metabolic control of the patients and that optimal control in most cases renders the abnormal Lp pattern towards normal. The insulin pump study clearly established that even within 2 weeks of optimal control, plasma glucose levels normalize, followed by a reduction in VLDL and an increase in HDL and ApoA. The abnormal atherosclerosis index improved significantly. There seem to exist only relatively few, mostly NIDD whose hyperlipoproteinemia is hardly controllable and not influenced by diet and oral antidiabetic drugs. These individuals most probably suffer from additional single or multiple genetic defects and have a latent form of hyperlipoproteinemia. If these individuals become diabetic, dietetic treatment may not be sufficient to normalize their Lp pattern. In such cases, hypolipidemic drugs, e.g. fibrates as shown in this study, have been found to influence beneficially the secondary hyperlipidemia. Frequently, a normalization could be observed with bezafibrate therapy (3 × 200 mg/day) within 2 months.

References

1 Alaupovic, P.: Apolipoproteins and lipoproteins. Atherosclerosis *13:* 141–146 (1971).
2 Kostner, G.M.: Apolipoproteins and lipoproteins of human plasma: significance for health and diseases. Adv. Lipid Res. *20:* 1–44 (1983).
3 Kostner, G.M.; Avogaro, P.; Cazzolato, G.; Marth, E.; Bittolo Bon, G.: Lipoprotein Lp(a) and the risk for myocardial infarction. Atherosclerosis *38:* 51–61 (1981).
4 Bittolo Bon, G.; Cazzolato, G.; Saccardi, M.; Kostner, G.M.; Avogaro, P.: Total plasma ApoE and high density lipoprotein ApoE in survivors of myocardial infarction. Atherosclerosis *53:* 69–75 (1984).
5 Schonfeld, G.; Birge, C.; Miller, P.; Kessler, G.; Santiago, J.: Apolipoprotein B levels and altered lipoprotein composition in diabetes. Diabetes *23:* 827–834 (1974).
6 Eckel, R.H.; Albers, J.J.; Cheung, M.C.; Lindgren, F.T.; Bierman, E.L.: High density lipoprotein composition in insulin-dependent diabetes mellitus. Diabetes *30:* 132–138 (1981).
7 Ewald, U.; Tuvemo, T.; Vessby, B.; Walinder, O.: Serum apolipoproteins AI, AII and B in diabetic children and matched healthy controls. Acta paediat. scand. *71:* 15–18 (1982).
8 Lopes-Virella, M.F.; Wohltmann, H.J.; Mayfield, R.K.; Loadholt, C.B.; Colwell, J.A.: Effect of metabolic control on lipid, lipoprotein, and Apo-Lp levels in 55 insulin-dependent diabetic patients. Diabetes *32:* 20–25 (1983).
9 Briones, E.R.; Mao, S.T.J.; Palumbo, P.J.; O'Fallon, W.M.; Chenoweth, W.; Kottke, B.A.: Analysis of plasma lipids and apolipoproteins in insulin-dependent and noninsulin-dependent diabetics. Metabolism *33:* 42–49 (1984).
10 Gabor, J.; Spain, M.; Kalant, N.: Composition of serum VLDL and HDL in diabetes. Clin. Chem. *2:* 1261–1265 (1980).
11 Stahlenhoef, A.F.H.; Demacker, P.N.M.; Lutterman, J.A.; vant'Laar, A.: Apolipoprotein C in type 2 diabetic patients with hypertriglyceridemia. Diabetologia *22:* 489–491 (1982).
12 Schernthaner, G.; Kostner, G.M.; Dieplinger, H.; Prager, R.; Mühlhauser, I.: Apolipoproteins AI, AII, B, Lp(a) and LCAT in diabetes mellitus. Atherosclerosis *49:* 277–293 (1983).
13 Pilger, E.; Pristautz, H.; Pfeiffer, K.H.; Kostner, G.M.: Retrospective evaluation of risk factors for peripheral atherosclerosis by stepwise discriminant analysis. Arteriosclerosis *3:* 57–63 (1983).
14 Dieplinger, H.; Kostner, G.M.: The determination of LCAT in the clinical laboratory: a modified enzymatic procedure. Clinica chim. Acta *106:* 319–325 (1980).
15 Prager, R.; Schernthaner, G.; Kostner, G.M.; Mühlhauser, I.; Zecher, R.; Dorda, W.: Effect of bezafibrate on plasma lipids, lipoproteins and Apo-Lp AI, AII, and

B and LCAT activity in hyperlipidemic noninsulin-dependent diabetics. Atherosclerosis *43:* 321–327 (1982).
16 Nestel, P.J.: Cholesterol metabolism in diabetes mellitus; in Waldhäusl, Proc. 10th. Congr. of the Int. Diabetes Fed., pp. 9–14 (Excerpta Medica, Amsterdam 1979).
17 Steinbrecher, U.P.; Witztum, J.L.: Glucosylation of LDL to an extent comparable to that seen in diabetes slows their catabolism. Diabetes *33:* 130–134 (1984).

Gerhard M. Kostner, MD, Institute of Medical Biochemistry, University of Graz, A-8010 Graz (Austria)

Genetic Determinants of Dyslipidemias

Jan L. Breslow

Rockefeller University, New York, N.Y., USA

Numerous studies have implicated altered levels of plasma lipoproteins in the pathogenesis of atherosclerosis. In particular, elevated LDL cholesterol and diminished HDL cholesterol appear to be strong risk factors for atherosclerosis susceptibility. Research of the last two decades has revealed a rather complex set of events that control plasma lipoprotein levels. Specific proteins have been implicated in the regulation of lipoprotein synthesis, interconversions, and catabolism. There are eight well-characterized apolipoproteins that coat the lipoprotein particles which carry cholesterol in the bloodstream. There are three or four other proteins that serve in the processes that control lipoprotein interconversions either as enzymes, cofactors, or transfer proteins. In addition, there are at least three well-characterized cell surface receptors that recognize lipoproteins and mediate their cellular uptake and catabolism. The techniques of molecular biology are now being applied to isolate the genes that code for these proteins. Studies of the structure and function of the normal genes are underway, and individuals with premature atherosclerosis are being examined for abnormalities in these genes. Once identified, these genes can be classified as atherosclerosis susceptibility genes and appropriate probes can be used to screen for them in presymptomatic individuals. Such individuals could be targeted for appropriate atherosclerosis prevention therapies.

At this time, cDNA and genomic clones have been isolated for all of the human apolipoproteins as well as the LDL receptor. This brief review will summarize our current knowledge of apolipoprotein gene structure, function and genetic variation. (A complete set of references is contained in references 1 and 2.)

It has been deduced from the cDNA sequences as well as cell-free synthesis experiments that the apolipoproteins are synthesized with N-terminal extensions. Signal peptides from 18 to 26 amino acids in length have been identified for Apo A-I, Apo A-II, Apo A-IV, Apo CI, Apo CII, Apo CIII, and Apo E. In addition, it has been found that Apo A-I and Apo A-II are synthesized with propeptides of 6 and 5 amino acids, respectively. The Apo A-I propeptide has a rather unusual structure in that it does not have two basic amino acids adjacent to the amino terminus of the mature protein; whereas Apo A-II has the typical propeptide structure. Although propeptides previously described for other proteins are cleaved intracellularly, it has been found that virtually all of Apo A-I and approximately half of Apo A-II are secreted as propeptides. This requires the existence of protease activities in lymph or plasma capable of cleaving these propeptides to the mature protein. Experiments are in progress to identify these proteases and study the physiological significance of these reactions.

Complete DNA sequences have been derived for three of the apolipoprotein genes, Apo A-I, Apo E and Apo CIII. These genes are compact and only 2–4 kb in length. They have a very similar structure, with each containing 4 exons and 3 introns. The introns are in strikingly similar locations with intron 1 in the 5′ untranslated region, intron 2 separating the signal peptide from the rest of the protein, and intron 3 interrupting the coding sequence of the mature protein between amino acids 40 and 61, thus separating the amino terminal amino acids from the carboxy terminal amino acids of the protein. Such intron locations suggest that they serve to separate exons that code for functionally distinct regions of the gene. In the cases of Apo A-I and Apo E, the 4th exon is elongated by 66 bp intragenic duplications. The consensus sequence for these duplications for the Apo A-I gene is 66% homologous with the consensus sequence of the duplications of the Apo E gene at the DNA level. Translation of the consensus sequence indicates that this region of the gene codes for amphipathic alpha helical segments presumably involved in lipid binding. A recent report of the cDNA sequence for rat Apo A-IV indicates many similar 66 bp intragenic duplication events. These presumably are in the Apo A-IV 4th exon. Thus, there seem to be great similarities between the apolipoprotein genes, suggesting common ancestral origin and that they are part of a multigene family.

Further experiments utilizing cDNA probes and somatic cell hybrids have localized the apolipoprotein genes to specific human chromosomes. Apo A-II has been localized to human chromosome 1. Apo A-I, Apo CIII and Apo A-IV all localize to human chromosome 11. Further sublocalization has been to the long arm of human chromosome 11. The genes for apolipoproteins E, CI and CII all localize to human chromosome 19. The LDL receptor gene is also on human chromosome 19, but is not thought to be closely linked to the apolipoprotein gene locus. The exact location and orientation of the apolipoprotein genes at the chromosome 11 and chromosome 19 loci have been investigated. The Apo A-I and Apo CIII genes have been shown to be within 2.5 kb of each other but reside in opposite orientations such that they share a common 3' untranslated region and are convergently transcribed. The human Apo A-IV gene has been shown to be closely linked to the A-I/CIII locus but the exact relationship is not yet defined. At the chromosome 19 locus, the Apo E and Apo CI genes have been shown to be within 4 kb of each other and in the same orientation. The Apo CII gene is closely linked to the Apo E/Apo CI locus but this exact relationship also has to be defined.

Genetic Variation

Human genetic variation has been defined in great detail for two of the apolipoproteins, Apo A-I and Apo E. Apo A-I is the principal structural protein in HDL and because of the importance of HDL levels in predicting atherosclerosis susceptibility, attention has been paid to possible Apo A-I genetic variation both by screening for variants in populations, principally by isoelectric focusing, as well as studying Apo A-I in individuals with altered HDL levels. Thus far, these studies have resulted in the discovery of eight proven Apo A-I variants. Seven of these variants have been shown in people who appear to be heterozygous for one normal Apo A-I structural allele and another allele that specifies a gene product that is either one charge unit more acidic or one charge unit more basic than wild-type. Approximately 1 in 400 to 1 in 500 individuals in the general population is heterozygous for one of these mutant Apo A-I forms. The specific amino acid substitutions responsible for these charge shifts have been elucidated (table I).

A unique Apo A-I gene lesion has been described in a family in which 2 sisters had very low HDL but normal LDL levels, xanthomas,

Table I. Apo A-I genetic variants

Name	Charge difference	Defect	HDL level
A-I$_{Milano}$	−1	Arg$_{173}$→Cys	↓
A-I$_{Marburg}$	−1	Lys$_{107}$→0	↓
A-I$_{Munster2}$	−1	Lys$_{107}$→0	NL
A-I$_{Giessen}$	+1	Pro$_{143}$→Arg	NL
A-I$_{Munster3A}$	+1	Asp$_{103}$→Asn	NL
A-I$_{Munster3B}$	+1	Pro$_4$→Arg	NL
A-I$_{Munster3C}$	+1	Pro$_3$→His	NL
A-I-CIII deficiency	NA	A-I gene insert	↓↓

severe premature atherosclerosis and absent plasma Apo A-I and Apo CIII. First-degree relatives of these individuals had half normal plasma levels of HDL, Apo A-I and Apo CIII. Southern blotting of genomic DNA from the probands, after digestion with EcoRI, with an Apo A-I cDNA probe revealed a single band of 6.5 kb; whereas normal genomic DNA showed a single 13 kb band. First-degree relatives, including the mother and father, of the probands showed one normal band and one abnormal band and appear to be carriers of a mutant allele associated with the Apo A-I gene. The probands appear to be homozygous for this mutant allele. Southern blotting of probands' DNA, after digestion with other restriction endonucleases, with an Apo A-I cDNA probe consistently revealed differences from wild-type. This suggested that the genetic lesion was not a single base pair substitution but rather a more major DNA alteration. Southern blotting with other probes derived from the region of the Apo A-I gene indicates that the 4th exon of the Apo A-I gene is interrupted at approximately the codon specifying residue 80 of the mature protein and this may explain the lack of Apo A-I in the plasma of these patients. Recently, the insertion sequences have been cloned and found to correspond to Apo CIII sequences. Thus, a DNA rearrangement at the Apo A-I, Apo CIII gene locus has occurred in these patients.

Genetic variation in Apo E has also been investigated extensively. Apo E participates in receptor-mediated lipoprotein uptake and is recognized both by the LDL or Apo B/E receptor as well as an Apo E receptor in liver that appears to be responsible for the clearance of chylomicron remnants. One-dimensional isoelectric focusing of human plasma Apo E reveals several bands whose relative concentrations vary between different individuals. Utilizing two-dimensional gel electrophoresis it was possible to determine that some of these bands were due to sialo-Apo E isoproteins and others due to variations in the isoelectric point of the major asialo-Apo E isoproteins. Studies of large numbers of individuals revealed six common Apo E phenotypes in the population. Family studies showed that these phenotypes were a result of a single Apo E gene locus with three common alleles. The alleles have been designated ε4, ε3 and ε2 and the gene products from basic to acidic are E4, E3 and E2, respectively. There are three homozygous phenotypes, E4/4, E3/3 and E2/2 and three heterozygous phenotypes, E4/3, E3/2 and E4/2. Five relatively large studies of Apo E phenotype prevalence have been reported. These have been done in diverse geographic areas, but primarily in Caucasians. The range of allele frequencies were ε4 11–15%, ε3 74–78% and ε2 8–13%. This common Apo E polymorphism has been found to play a role in type III hyperlipoproteinemia. This disorder is characterized by elevated cholesterol and triglyceride levels, as as result of delayed chylomicron remnant clearance, xanthomas, and premature coronary as well as peripheral vascular disease. Over 90% of individuals with type III hyperlipoproteinemia have the E2/2 phenotype; whereas this occurs in only 0.5–1.4% of normal individuals. In addition, when E2 is isolated and studied in vitro, it does not bind as well as E3 or E4 to high affinity lipoprotein receptors. It has been suggested that chylomicron remnants with E2 on their surface are recognized poorly by receptors, cleared slowly, and accumulate in plasma. Chylomicron remnants are quite potent stimulators of macrophage cholesteryl ester accumulation in vitro and high plasma concentrations of these particles may be involved in the atherogenic process in vivo. These data all suggest that homozygosity for the ε2 allele may be the underlying cause of type III hyperlipoproteinemia. However, the disease frequency is such that only 1–2% of people with the E2/2 phenotype actually express the disease. It is known that other hormonal and environmental factors are necessary for disease expression. However, the current belief is that type III hyperlipoproteinemia is a result of two gene defects, one of these is the Apo E structural gene and the other in a gene that influences chylomicron

Table II. Human Apo E protein polymorphism

Name	Charge difference	Defect
E7	+4	?
E5	+2	?
E4	+1	$Cys_{112} \to Arg$
E3	0	–
E3*	0	$Ala_{99} \to Thr, Ala_{152} \to Pro$
E3**	0	$Cys_{112} \to Arg, Arg_{142} \to Cys$
E2	−1	$Arg_{158} \to Cys$
E2*	−1	$Arg_{145} \to Cys$
E2**	−1	$Lys_{146} \to Gln$
E1	−2	$Gly_{127} \to Asp, Arg_{158} \to Cys$

remnant synthesis or catabolism in a synergistic fashion. The second gene product is yet to be identified. In addition to the striking involvement of the E2/2 phenotype in type III hyperlipoproteinemia, it appears that the Apo E gene locus may be one of the factors influencing lipid levels in the general population. A recent review of five studies in the literature suggest that the ε2 allele exerts a stepwise gene dosage effect on lowering LDL cholesterol levels. The ε2 allele also appears to influence the VLDL fraction and results in an increase in VLDL cholesterol and triglyceride levels in a similar stepwise manner. Recent studies have found E2 more frequent in patients with hypertriglyceridemia, E4 more frequent in hypercholesterolemia, and E2 and E4 are more frequent in mixed hyperlipidemia and suggest a specific effect of E4 on blood lipid values.

Amino acid sequence analysis established that the two common variants of Apo E, E4 and E2, differ from E3 by a single amino acid substitution. E4 differs from E3 at residue 112 because of an arginine for cysteine substitution and E2 differs at residue 158 because of a cysteine for arginine substitution. Isoelectric focusing and amino acid and DNA sequencing have identified other rare Apo E alleles. In all, at least 10 Apo E alleles are known and these are listed. In all but two of these instances, E7 and E5, the amino acid substitution underlying the variation has been identified (table II).

References

1 Breslow, J.L.: Human apolipoprotein molecular biology and genetic variation; in Richardson, Annual review of biochemistry, No. 54, pp. 699–727 (Annual Reviews, Inc., California 1985).
2 Zannis, V.I.; Breslow, J.L.: Genetic mutation affecting human lipoprotein metabolism; in Hirschhorn, Advances in human genetics, No. 14, pp. 125–215 (Plenum Press, New York 1985).

Jan L. Breslow, MD, The Rockefeller University, 1230 York Avenue, New York, NY 10021 (USA)

Genetic Polymorphisms in the Analysis of Diabetes mellitus[1]

David J. Galton

Medical Professorial Unit, St. Bartholomew's Hospital, London, UK

Introduction

A major goal in the prevention of diabetes is the ability to predict which individuals within a pedigree will develop the disease. Appropriate modification of nutritional and other environmental factors from an early age may then delay onset of the disease and its complications. Accurate prediction of susceptible individuals will require a detailed analysis of the genetic components that contribute to the disease phenotype [1]. Hitherto, attempts at a genetic analysis have been mainly indirect by studying protein markers or variants that associate with the disease or segregate with affected members of a pedigree. Such an approach cannot identify regulatory gene variants that do not result in the production of cytoplasmic protein; and cannot take into account the great amount of nucleotide variation in the human genome that may be influenced by natural selection but is not necessarily reflected by a mutant protein. With the advent of recombinant DNA technology, it is now possible to isolate human genes from any nucleated cell, transfer such genes into bacteria for cloning, and positively identify the nature of the gene by rapid DNA sequencing techniques. The clinician is then in a position to use such genomic probes for a direct analysis of the contribution of genetic variants to the manifestation of the disease and possibly allow prediction of susceptible individuals. Genomic probes can be used for at least four types of genetic studies:

---

[1] This work was supported by grants from the Medical Research Council, British Diabetic Association and the Fritz-Thyssen Foundation to which the author is very grateful.

Table 1. Genomic probes for the analysis of diabetes mellitus

Disease	Protein product		Gene
Type 1 (insulin-dependent) diabetes	HLA antigens complement proteins immune response proteins		HLA genes complement genes –
Type 2 (noninsulin-dependent) diabetes	hormones:	insulin insulin growth factors	insulin IGF I and II
	receptors:	insulin receptor	
	postreceptor enzymes:	hexokinase phosphofructokinase pyruvate kinase lipoprotein lipase	– – – –

(1) A direct study of the fine structure of any gene by restriction enzyme mapping where an abnormal gene product is suspected to influence the aetiology of the disease. Good examples are the insulin gene mutants in some types of noninsulin-dependent diabetes mellitus producing an insulin with impaired biological activity [2, 3].

(2) A study of DNA polymorphic sites adjacent to the relevant gene that may either affect some aspect of regulation or gene expression; or provide a marker for the affected gene to allow its identification in affected family members in a pedigree analysis [4]. A linked polymorphism can be used in tracing a disease-specific gene throughout members of an affected pedigree [5].

(3) A study of DNA polymorphic sites adjacent to the relevant gene that may be in linkage disequilibrium with other, hitherto unsuspected disease-specific genes close to this locus, and may help in their eventual identification.

(4) To study the arrangement of gene clusters, where genes involved in a particular aspect of metabolism may occur closely together on the same chromosome. Good examples of this are the β-globin gene cluster on chromosome 11 involved in oxygen transport; and the apoprotein A1/C-III gene cluster involved in lipid transport [6].

The genetic variants that may possibly be involved in conferring predisposition to diabetes mellitus are listed in table I. Obvious candidate genes are the ones coding for proteins considered to be involved in the pathogenesis of diabetes mellitus. Genes coding for insulin, the insulin receptor and intracellular enzymes underlying postreceptor defects [7] may all be involved in the pathogenesis of type II diabetes; whereas genes coding for the HLA-D loci, complement protein loci and other immune response genes may be involved in type I diabetes. Other genes that do not directly code for cellular proteins may also be involved.

This chapter considers genetic variants related to the human insulin gene since it was the first gene to be used in the genetic analysis of diabetes mellitus. It is very likely that the other genes in table I will soon become available to extend such studies.

Insulin Gene and Diabetes-Related Disorders

Recent advances in recombinant DNA methods now permit a direct analysis of the insulin gene and other genes that may contribute to the genetics of diabetes and its related disorders. The insulin gene was first isolated, cloned and sequenced by *Bell* et al. [8] in 1980. Messenger RNA was extracted from a human insulinoma (a large proportion of which codes for insulin); and then using a rat cDNA gene probe they isolated mRNA for human insulin, and hence obtained a cDNA clone for the hormone using reverse transcriptase. The latter clone was then used as a probe to isolate the human insulin gene from a genomic library.

Insulin Gene Structure

The human insulin gene, located on the short arm of chromosome 11 [9] is 1,355 base pairs in length, comprising three coding regions (or exons) separated by two introns (fig. 1). The first exon contains DNA sequences specifying ribosomal binding sites on the final mRNA product. Within the second exon is the start codon (ATG) followed by sequences coding for the signal peptide, B chain and C peptide. The DNA sequence coding the C peptide is interrupted by a second intron. The sequence for the A chain is found in the third exon. At the 3'-end of the gene the site of the polyadenine (poly-A) tail is located 74 nucleotides from the stop codon.

Regulatory sequences are found in the 5'-flanking region. The promotor region (or TATAA sequence) which binds RNA polymerase

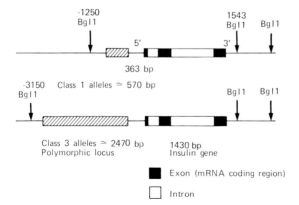

Fig. 1. A map of the human insulin gene located on chromosome 11. Note the large insertions (hatched areas) at the 5'-end of the gene.

is located 25 base pairs from exon 1; and a gene enhancer sequence may occur between bases 168–258 from exon 1.

The first product of the insulin gene, or its primary transcript, contains all the transcribed sequence from the first nucleotide of exon 1 to the site of the poly-A tail on the 3'-side of the gene, including the introns. The next step involves enzymatic splicing out of the two introns, leaving the mature mRNA. Subsequently, the mRNA is translated into the peptide product of the insulin gene, pre-proinsulin, and is available for storage and secretion.

DNA Variation In and Around the Insulin Gene

Within the Coding Sequence. Insulin is one of the most closely conserved biological molecules in nature and this is reflected in the conservation of the amino acid sequence in phylogenetically older species. For example, hagfish insulin has 80% homology with human insulin. At the DNA level only a few mammals have been shown to have a different gene structure. Rat, mouse and several fish species have been shown to possess two separate nonallelic insulin genes. The rat insulin gene 1 differs from gene 2 by not having the first intron. Rat insulin gene 2 is almost identical in DNA sequence to the human insulin gene, possessing three exons and two introns.

Within the human insulin gene there is a certain amount of allelic variation. Of four genes that have been sequenced, they differed from each other by four nucleotides. The positions of these polymorphisms were one within intron 1; one within intron 2; and two in the 3'-untranslated region. They appeared to be without pathological effects [10].

DNA Insertions in the 5'-Flanking Region of the Insulin Gene. In contrast to the close homology found within the insulin gene locus amongst animal species, there is a highly polymorphic region of DNA beginning 363 base pairs from the start of insulin mRNA synthesis (the hatched box of figure 1). This region can be broadly subdivided into alleles containing one of three DNA inserts of variable length: short DNA insertions, 0–600 base pairs (class 1 allele); intermediate sized insertions, 600–1,600 base pairs (class 2 allele), and long DNA insertions of 1,600–2,200 base pairs (class 3 allele) [11]. The nucleotide structure of these three insertions are very similar, being composed of a variable number of a 14 base pair oligonucleotide of consensus sequence, ACAGGGGTGTGGGG [15].

Hence the class 1, 2 and 3 alleles differ only in possessing on average 40, 80 or 160 tandem duplications of this oligonucleotide, respectively. Although it is likely from nucleotide sequence data that nearly every individual may have slight differences in the total nucleotide sequence of the insertion class, for the purpose of clinical studies, it is still not possible to make use of this information. For such studies restriction enzyme analysis is used, which even in the most capable hands can only detect differences of 50 base pairs in length between allelic variants. Man, being diploid, inherits an insulin gene allele from both parents and hence the following genotypes can be observed: 1/1 (homozygous small); 1/2, 1/3 (heterozygous for large and small), and 2/2, 2/3 and 3/3 (homozygous large). This region can therefore be used as a genetic marker for the two parental insulin genes as more than 63% of individuals are heterozygous with respect to length at this point. In practice, in Caucasian subjects the class 2 allele is uncommon and for this reason most studies have combined data of the class 2 with the class 3 allele. The bimodal distribution of DNA inserts in Caucasian subjects can be seen in figure 2 and this is contrasted with the different distributions found in other racial populations. For instance, in American blacks a trimodal distribution is found with a higher frequency of the class 2 allele, and in Asian populations a decreased frequency of the class 3 allele is found compared to Caucasian subjects [12].

This polymorphic region is of great interest as its nucleotide sequence is unique to the human genome. Together with its proximity

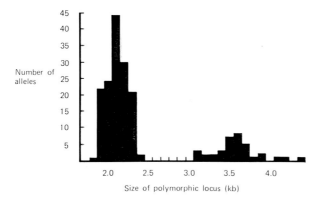

Fig. 2. Histogram of sizes of insulin gene fragments containing class 1 and class 3 DNA insertions. (Rsa 1 digestion, 180 alleles.)

to the putative enhancer locus [13] and the non-Gaussian distribution of inserts in this region, it has been postulated that this region may be more than a genetic marker and may be a region of functional importance for the regulation of insulin gene transcription.

Only one other example has been reported of an oligonucleotide repeat region similar to the insulin gene polymorphism. It is found in the α-globin cluster between the embryonic haemoglobin genes, zeta 1 and zeta 2, and 3' to the α-globin gene [14]. The oligonucleotide repeat between the zeta 1 and 2 genes has been sequenced and its consensus sequence only differs by three base pairs from the insulin polymorphic repeat.

Direct evidence implicating the insulin gene polymorphism in transcription of the insulin gene is as yet lacking. Microinjection techniques using intracellular transcription systems have failed to detect any differences in insulin gene function, whether it possesses a large or short insertion of DNA. Another possibility that the region codes for a protein perhaps functioning like a prokaryote repressor or inducer is unlikely. From sequence data every third amino acid would be glycine or proline depending on which strand is transcribed.

Until there is evidence to the contrary, this polymorphism should be considered as a good genetic marker for the two parental insulin genes and as such can be used in clinical studies. The aim then is to seek for disease associations with the different insulin gene variants.

Table II. Genotype distribution of the polymorphic locus adjacent to the insulin gene in type II diabetes mellitus: controls are significantly different from diabetics in the studies of *Rotwein* et al. [16] and *Hitman* et al. [18]

Study group	Genotype distribution in:			
	controls		type II diabetes	
	homozygous class 1 allele (1/1)	homozygous class 3 allele (3/3)	homozygous class 1 allele (1/1)	homozygous class 3 allele (3/3)
Bell et al [15]	13	2	6	1
Rotwein et al. [16]	45	3	43	11
Owerbach and *Nerup* [17]	31	4	22	8
Hitman et al [18]	37	7	19	20

Disease Associations Found With the Insulin Gene Polymorphic Region

Type II or Noninsulin-Dependent Diabetes

Prevalence Studies. Four centres have sought for disease associations with the polymorphic region adjacent to the insulin gene, and the results are summarized in table II. The most common genotype found in Caucasian populations is heterozygous for the class 1 and 3 alleles (1/3), accounting for 50% of subjects studied; 42% of subjects had the genotype homozygous for the class 1 allele (1/1), and the rarest genotype was homozygous for the class 3 allele (3/3), being present in less than 10% of the population. Another approach is to examine the frequencies for the class 1 and 3 alleles based on the Hardy-Weinberg principle. If the frequency of the class 1 allele is p and the frequency of class 3 allele is q (and p + q = 1), then the genotype frequencies can be calculated from the allele frequencies by the formula:

$p^2:2pq:q^2$

This assumes random mating conditions, Mendelian inheritance and no disturbing conditions of mutation, selection or migration. The allele frequencies of the class 1 allele was approximately 0.67 and class

3 allele 0.33 in control populations from London, St. Louis, San Francisco and Copenhagen. In type II diabetics a greater preponderance of the genotype homozygous for the class 3 allele has been reported by some groups with an increased relative incidence of about 5. An even greater preponderance of the homozygous class 3 genotype has been found in a subgroup of diabetics who have coexisting hypertriglyceridaemia (relative incidence of 10.5).

Pedigree Analysis. An alternative approach to looking for disease associations is to examine for linkage of the polymorphic inserts with diabetes in large pedigrees. In a large pedigree of maturity-onset diabetes of the young, 17 diabetics and 38 nondiabetics, *Owerbach* et al. [19] found no close linkage between DNA inserts flanking the insulin gene and diabetes. Smaller MODY pedigrees have also been studied with no apparent association being found, although they were too small to be statistically analyzed. Additionally, our laboratory studied a large family (11 diabetics and 56 nondiabetics) more typical of type II rather than MODY (because of later age of onset of diabetes, 30 ± 10 years, and the presence of diabetic complications). Again, no close linkage was found of the DNA inserts in the 5'-flanking region of the insulin gene with diabetes (fig. 3). In conclusion, in the pedigrees so far analyzed, confirmation of genetic linkage of any DNA insert near the insulin gene with type II diabetes has not been found.

There are, however, problems with pedigree analysis in type II diabetes. Firstly, the age of onset of the disease is very variable, being more common in the 4th and 5th decades. As one cannot confidently predict who might become diabetic, this makes statistical analysis difficult when younger members are included. Secondly, as diabetic complications lead to premature death (from myocardial infarction, renal failure), key older members of the family are sometimes not available for genotyping.

Type I or Insulin-Dependent Diabetes

Three centres have studied a total of 169 insulin-dependent diabetics with the insulin gene probe searching for disease associations with the polymorphic locus. In the largest Caucasian study from San Francisco [20], type I diabetics were compared to 83 controls. The genotype frequencies in type I diabetics showed a greater preponderance of the homozygous class 1 insertions (76%) compared to controls of 44%.

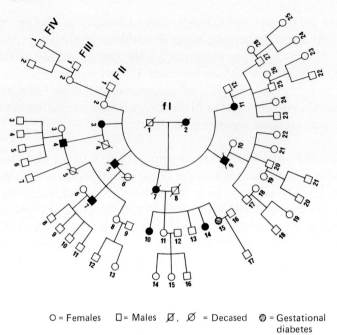

O = Females   □ = Males   Ø, Ø = Deceased   ⊙ = Gestational diabetes

Fig. 3. Analysis of a large pedigree for linkage of diabetes with the polymorphic locus close to the insulin gene. The family members were genotyped using the hypervariable locus adjacent to the insulin gene and analyzed for linkage by Lod scores with an age correction and assuming an autosomal-dominant mode of inheritance. Results of analysis excluded any close linkage between genotype and disease (Lod scores were −5.72 and −1.125 for recombination fractions of 0 and 0.1, respectively).

Table III. Genotype distribution of the polymorphic locus adjacent to the insulin gene in type I diabetes: genotype distributions between diabetics and controls are highly significant when tested by chi-square

Study group	n	Genotype distribution in:			reference
		homozygous class 1 allele (1/1)	heterozygous (1/3)	homozygous class 3 allele (3/3)	
Controls	88	37 (42%)	44 (50%)	7 (8%)	
Type I diabetes	53	42 (79%)	11 (21%)	0 (0%)	21
Type I diabetes	113	86 (76%)	27 (24%)	0 (0%)	20

Other studies have confirmed these results and are summarized in table III. The strength of the association of type I diabetics with the genotype 1/1 is stronger than that of type II diabetics with the genotype 3/3. Results for the Lod score analysis of type I pedigrees excludes any close linkage between genotype 1/1 and disease.

Significance of Genotype Distributions

The particular distribution of genotypes in populations found for this polymorphic locus adjacent to the insulin gene could have arisen by random genetic drift and the associations found with the various disease states be entirely spurious. However, there are several facts against this interpretation: (1) the nonrandom distribution of DNA insertional elements into two (or three) defined classes (fig. 1) suggest that selective forces may be operating and that intermediate-sized insertions are in some way disadvantageous; (2) the homozygous large insertional class (3/3) is present at low frequencies in the population and shows a clear disease association with glucose and triglyceride intolerance; but (3) the same does not apply for the homozygous short insertions (1/1) which are present at a frequency of greater than 40% in the healthy population and yet also show an association with a disease known to affect fertility and fecundity (type I diabetes mellitus).

It is impossible, however, from the presently available data to account for the stage of evolution of these different genotypes. If one considers the past evolutionary history of these genotypes, the homozygous short (1/1) class may be in balance with the other genotypes; or be gradually increasing or decreasing in frequency. If the homozygous short genotype were increasing it would be very difficult to interpret its association with type I diabetes; unless there were a subgroup of individuals with the homozygous short genotype who have an additional mutation in linkage disequilibrium that confers susceptibility to type I diabetes.

Assuming that natural selection is operating on this polymorphic locus, there are several interesting possibilities:

(1) The polymorphic locus is so close to the regulatory sequences on the 5'-flanking region of the insulin gene that some sized insertions may affect insulin gene transcription, leading to inappropriate rates of insulin synthesis. No differences to support this have so far been found

in the first phase of insulin secretion between patients with genotype 1/1 as compared to 1/3 + 3/3. Alternatively, the highly variable region of DNA may be acting as a genetic marker for an abnormality within the coding region for insulin producing a mutant insulin. This would be analogous to the Hpa 1 polymorphism 3′ to the β-globin gene that associates with the sickle cell mutation. Such a mutant insulin has not so far been found.

(2) Some classes of DNA insertions at the polymorphic locus may be in linkage disequilibrium with other disease-specific genes on the short arm of chromosome 11. A model for this interpretation would be the studies associating type I diabetes with HLA-related antigens on chromosome 6. Initially an association with the HLA antigens B8 and B15 were found, but stronger associations have been subsequently found with HLA-DR3 and -DR4, respectively. In a similar manner the polymorphic locus on chromosome 11 may be associated with other alleles that determine susceptibility to diabetes, as yet unidentified.

(3) The variable disease associations found with this polymorphic locus may be a manifestation of a polygenic disease; where other genes are required to interact before predisposition to diabetes become apparent. This is very likely to be the case with diabetes where it is known that heterogeneity exists within both types I and II. Within each subset two or more disease susceptibility genes may be required for interaction with environmental factors before manifestation of the disease.

References

1 Galton, D.J.: Molecular genetics of common metabolic disease (Arnold, London 1985).
2 Tager, H.; Given, B.; Baldwin, D.; Mako, M.; Markose, J.; Rubenstein, A.; Olefsky, J.; Kobayashi, M.; Kolterman, O.; Poucher, R.: A structurally abnormal insulin causing human diabetes. Nature, Lond. *281:* 122–125 (1979).
3 Kwok, S.C.M.; Steiner, D.F.; Rubenstein, A.H.; Tager, H.S.: Identification of a point mutation in the human insulin gene giving rise to a structurally abnormal insulin. Diabetes *32:* 872–875 (1983).
4 Rees, A.; Shoulders, C.C.; Stocks, J.; Galton, D.J.; Baralle, F.E.: DNA polymorphism adjacent to the human apoprotein A-1 gene: relation to hypertriglyceridaemia. Lancet *i:* 444–446 (1983).
5 Humphries, S.E.; Williams, L.; Myklebost, O.; Stalenhoef, A.F.H.; Demacker, P.N.M.; Baggio, G.; Crepaldi, G.; Galton, D.J.; Williamson, R.: Familial apolipoprotein C-II deficiency: a preliminary analysis of the gene defect in two independent families. Hum. Genet. *66:* 151–155 (1984).

6   Karathanasis, S.K.; Zannis, V.I.; Breslow, J.L.: Linkage of human apolipoproteins A-I and C-III genes. Nature, Lond. *304:* 371–372 (1983).
7   Taylor, K.G.; Holdsworth, G.; Galton, D.J.: Insulin independent diabetes: a defect in the activity of lipoprotein lipase in adipose tissue. Diabetologia *16:* 313–317 (1979).
8   Bell, G.I.; Pictet, P.L.; Rutter, W.J.; Cordell, B.; Tischer, E.; Goodman, H.M.: Sequence of the human insulin gene. Nature, Lond. *284:* 26–32 (1980).
9   Owerbach, D.; Bell, G.I.; Rutter, W.J.; Brown, J.A.; Shows, T.B.: The insulin gene is located on the short arm of chromosome 11. Diabetes *30:* 267–270 (1981).
10 Ullrich, A.; Dull, T.J.; Gray, A.; Brosius, J.; Surer, I.: Genetic variation in the human insulin gene. Science *209:* 612–614 (1980).
11 Bell, G.I.; Karam, J.H.; Rutter, W.J.: Polymorphic DNA region adjacent to the 5′-end of the human insulin gene. Proc. natn. Acad. Sci. USA *78:* 5758–5766 (1981).
12 Williams, L.G.; Jowett, N.I.; Vella, M.; Humphries, S.E.; Galton, D.J.: Allelic variation adjacent to the human insulin and apolipoprotein C-II genes in different ethnic groups. Hum. Genet. (submitted).
12 Walker, M.D.; Edmund, T.; Boulet, A.M.; Rutter, W.J.: Cell specific expression controlled by the 5′-flanking region of the insulin and chymotrypsin genes. Nature, Lond. *306:* 557–561 (1983).
14 Higgs, D.R.; Goodbourn, S.E.Y.; Wainscot, J.S.; Clegg, J.B.; Weatherall, D.J.: Highly variable region of DNA flanking the human α-globin genes. Nucl. Acids Res. *9:* 4213–4219 (1981).
15 Bell, G.I.; Karam, J.H.; Rutter, W.J.: Polymorphic DNA region adjacent to the 5′ end of the human insulin gene. Proc. natn. Acad. Sci. USA *78:* 5758–5766 (1981).
16 Rotwein, P.S.; Chirgwin, J.; Province, M.; Knowler, W.C.; Pettit, O.J.; Cordell, B.; Goodman, H.M.; Permutt, M.A.: Polymorphism in the 5′-flanking region of the human insulin gene: a genetic marker for non-insulin dependent diabetes. New Engl. J. Med. *308:* 65–71 (1983).
17 Owerbach, D.; Nerup, J.: Restriction fragment length polymorphism of the insulin gene in diabetes mellitus. Diabetes *31:* 275–277 (1982).
18 Hitman, G.A.; Jowett, N.I.; Williams, L.G.; Humphries, S.E.; Winter, R.M.; Galton, D.J.: Polymorphism in the 5′-flanking region of the insulin gene and non-insulin dependent diabetes. Clin. Sci. *66:* 383–388 (1984).
19 Owerbach, D.; Thomsen, B.; Johansen, K.; Lamm, L.U.; Nerup, J.: DNA insertion sequences near the insulin gene are not associated with maturity-onset diabetes of young people. Diabetologia *25:* 18–20 (1983).
20 Bell, G.I.; Horita, S.; Karam, J.H.: A highly polymorphic locus near the human insulin gene is associated with insulin-dependent diabetes mellitus. Diabetes *33:* 176–183 (1984).
21 Hitman, G.A.; Tarn, A.C.; Winter, R.M.; Drummond, V.; Williams, L.G.; Jowett, N.I.; Bottazzo, G.F.; Galton, D.J.: Insulin-dependent diabetes associates with a highly variable locus close to the insulin gene. Diabetologia (in press, 1985).

David J. Galton, MD, Medical Professorial Unit, St. Bartholomew's Hospital, London EC1 (UK)

Very Low Density Lipoprotein Triglyceride Metabolism in Diabetes[1]

Esko A. Nikkilä

Third Department of Medicine, University of Helsinki Hospital, Helsinki, Finland

It is generally recognized that the lipid and lipoprotein concentrations of diabetic patients are highly variable. This is partly due to the heterogeneity of the disease and partly due to the wide variety of other factors associated with diabetes. These include, for example, the degree of glycemic control, insulin sensitivity of the patient, insulin dosage and method of administration, amount of body fat and the presence or absence of diabetic renal disease. Moreover, all the genetic or exogenous factors influencing plasma lipoprotein concentrations in nondiabetic individuals will evidently be operative also in diabetic patients.

Of various plasma lipoproteins, very low density lipoprotein (VLDL) represents the fraction which is probably most sensitively changed in diabetes. This may be understood on the basis of the regulatory function of insulin in both production and removal of plasma VLDL triglycerides. Plasma insulin concentration and the insulin sensitivity of various tissues influence at least the rate of release of free fatty acids (FFA) from fat cells (lipolysis) and the rate of catabolism of triglyceride-rich lipoproteins (VLDL and chylomicrons). The flux of FFA into the liver is evidently the major regulatory factor of VLDL production but insulin may have also direct effects on the intrahepatic processes responsible for the rate of VLDL secretion. Published data on this latter effect are highly controversial, however. In perfused rat liver addition of insulin has been found to increase VLDL output [*Topping and Mayes,* 1972] or to be without effect [*Nikkilä,* 1974]. In isolated rat hepatocytes both stimulation [*Beynen* et al., 1981] and inhibition [*Durrington* et al., 1982; *Patsch* et al., 1983] of

---

[1] Supported by grants from the State Medical Research Council (Academy of Finland).

VLDL output has been observed upon addition of insulin to the medium. In in vivo studies in rats a chronic insulin administration has been found to stimulate the production of VLDL [*Steiner* et al., 1984]. In man an acute intraportal infusion of insulin may also increase the hepatic output of VLDL in spite of a marked reduction of FFA uptake [*Vogelberg* et al., 1980]. In insulin-deficient rat liver the hepatic triglyceride secretion rate at any given perfusate FFA level is less than in respective normal liver [*Reaven and Mondon,* 1984]. If any conclusions can be drawn from these results it would be that insulin probably stimulates the VLDL synthesis and secretion by its direct hepatic action but, on the other hand, it inhibits this process by reducing the substrate flux. The net result is a complex sum of these opposing processes.

The effects of insulin on the catabolism of chylomicrons or VLDL are also complex. It is likely that the tissues responsible for the major part of VLDL degradation are skeletal muscle, adipose tissue and heart which show the highest activity of lipoprotein lipase. This enzyme is dependent on insulin as shown by an increase of its activity following oral glucose [*Goldberg* et al., 1980; *Taskinen and Nikkilä,* 1981] or insulin infusion during euglycemic clamp [*Sadur and Eckel,* 1982; *Yki-Järvinen* et al., 1984]. Thus, in principle, hyperinsulinemia enhances the activity of lipoprotein lipase and accelerates the catabolism of VLDL. Yet, the insulin resistance combined with hyperinsulinemia may neutralize these effects or even lead to a state where the cells are short of insulin in spite of excessive plasma insulin concentrations. This may be the situation particularly in type 2 diabetes where severe insulin resistance is associated with relative insulin deficiency.

Against this background it may be easily understood that the abnormalities of VLDL metabolism in different types of diabetes and in individual diabetic patients are far from uniform. Accordingly, the concentration of VLDL particles in diabetes ranges from low normal to grossly elevated levels. Moreover, there are also changes in the composition and in subclass particle distributions within the whole spectrum of VLDL. The subject has been recently reviewed in detail [*Nikkilä,* 1984].

Untreated Type 1 Diabetes (Insulin Deficiency)

A total or subtotal insulin deficiency is a rare condition in man. Indeed, it is observed only in previously untreated newly detected juvenile diabetes or in insulin-dependent diabetic patients who deliberately stop insulin administration. Yet, ketoacidosis may develop also in the presence of insulin when the antagonist hormone levels rise. Almost invariably a severe or even moderate ketosis is associated with raising VLDL concentrations (S_f 20–100 and 100–400 subclasses). Some patients even develop fasting hyperchylomicronemia with turbid appearance of plasma and presence of eruptive skin xanthomatosis [*Kolb* et al., 1955; *Weidman* et al., 1982]. In general, the values fall rapidly after initiation of insulin treatment but in some subjects clearly elevated triglyceride and VLDL levels may persist for days or weeks. The composition of VLDL is not much altered after initiation of insulin treatment which suggests that the VLDL particles accumulated during insulin deficiency do not much differ from respective lipoproteins of nondiabetic subjects.

The kinetic background of the VLDL elevation in insulin deficiency is not well established. In fact, none of the currently available methods for measuring either triglyceride or apoprotein kinetics of VLDL particles is fully reliable in the presence of marked elevation of VLDL levels. This failure is due to the heterogeneity of VLDL particle distribution and saturation of the removal system by the much enlarged VLDL pool size. Endogenous glycerol labeling and application of single-pool model for analysis of the triglyceride kinetics has shown that the production of VLDL triglycerides is markedly increased in diabetic ketoacidosis and also in nonketotic but poorly controlled type 1 diabetes [*Nikkilä and Kekki*, 1973]. The overproduction leads easily to an increase of plasma VLDL concentration since the removal capacity is restricted in the absence of insulin [*Nikkilä and Kekki*, 1973] due to subnormal lipoprotein lipase activity [*Nikkilä* et al., 1977; *Taskinen and Nikkilä*, 1979]. Thus, the available human data suggest that the elevation of VLDL particle concentration and of plasma triglyceride levels in uncompensated insulin deficiency result from a combination of increased secretion and deficient catabolism.

Insulin-Treated Diabetes

During conventional insulin treatment most type 1 diabetic patients have fairly normal VLDL and triglyceride concentrations. They may even show a tendency to be at the low side of normal range

[*Nikkilä and Hormila*, 1978]. However, with impairment of the glycemic control the VLDL increases and is positively correlated with blood glucose or hemoglobin A_1 [*Sosenko* et al., 1980; *Lopes-Virella* et al., 1981]. For some unknown reason, diabetic women are more sensitive to the hypertriglyceridemic effect of hyperglycemia than diabetic men [*Walden* et al., 1984]. Patients brought to strict glycemic control by either conventional insulin injections [*Lopes-Virella* et al., 1983] or by the use of insulin pump [*Petri* et al., 1983] have subnormal levels of VLDL and triglycerides.

The kinetics of VLDL metabolism are also nearly normal in insulin-treated diabetic patients [*Nikkilä and Kekki*, 1973; *Greenfield* et al., 1980; *Pietri* et al., 1983]. Both the VLDL triglyceride synthetic rate and the fractional catabolic rate are usually within normal range apart from patients with poor glycemic control who produce the VLDL at double normal rate [*Nikkilä and Kekki*, 1973]. In patients kept euglycemic by continuous insulin infusion the VLDL triglyceride production is suppressed to subnormal levels but the fractional catabolic rate remains normal [*Pietri* et al., 1983]. This explains the low plasma VLDL concentration in the well-controlled insulin-treated diabetic patients.

The lipoprotein lipase activity as well as the removal of exogenous fat (Intralipid) are either normal or at high side of normal range [*Nikkilä* et al., 1977; *Nikkilä and Hormila*, 1978]. The clearance of Intralipid increases upon insulin treatment [*Lewis* et al., 1972].

Type 2 Diabetes

Hypertriglyceridemia with elevated VLDL levels is frequently present in patients with noninsulin-dependent diabetes. This may be understood since both obesity and diabetes have separate effects on VLDL metabolism. The degree of hypertriglyceridemia is highly variable as is also the severity of hyperglycemia. Some patients may show a gross elevation of VLDL and fasting chylomicronemia (type 5 pattern) in the presence of mild diabetes while others have only slight elevation of plasma triglycerides in spite of a severe disturbance of glucose metabolism. Completely normal VLDL levels have been monitored in many nonobese type 2 diabetic patients. Diabetes may coincide with familial hypertriglyceridemia causing a 'diabetic-hyperlipemic syndrome' [*Adlersberg and Wang*, 1955; *Nikkilä and Kekki*, 1973; *Chait*

Table I. Production rate of VLDL triglycerides (mg/h/kg b.w.) in type 2 diabetes

Diabetic patients			Controls	Reference
obese	nonobese[1]	hyper-lipidemic		
14.1	13.3	16.2	7.8[2]	*Nikkilä* and *Kekki*, 1973
26.3[3]	15.0[3]	–	14.1[2]	*Kissebah* et al., 1974
14.7[3]	–	–	7.6[2]	*Greenfield* et al., 1980
–	27.4[3]	–	12.1	*Ginsberg* and *Grundy*, 1982
–	23.0	20.9	14.6[4]	*Kissebah* et al., 1982
–	14.0	21.2	11.4	*Abrams* et al., 1982
8.7	–	–	7.2[4]	*Howard* et al., 1983
–	–	28.5	10.0[4]	*Dunn* et al., 1984

[1] Includes also subjects with mild obesity.
[2] Controls nonobese.
[3] Includes both normolipidemic and hypertriglyceridemic subjects.
[4] Controls weight-matched.

et al., 1981] where plasma is of milky appearance and the patients present with eruptive xanthomatosis, pancreatitis or coronary heart disease.

The composition of VLDL particles remains fairly normal in diabetes [*Howard* et al., 1978; *Kissebah* et al., 1982] suggesting that the elevated triglyceride levels are accounted for by an increased VLDL particle count rather than by the presence of triglyceride-enriched VLDL particles. Several kinetic studies have almost uniformly shown that the production rate of VLDL triglycerides and VLDL Apo B is increased in patients with type 2 diabetes as illustrated by table I. This is particularly the case in diabetic patients with elevated triglyceride and VLDL levels but in many studies also the diabetic patients who have normal or near normal triglyceride concentrations and who are not particularly obese show higher average VLDL triglyceride synthetic rates than nondiabetic subjects. On the other hand, there are many diabetic patients who in spite of at least moderate hyperglycemia have no detectable abnormality in VLDL kinetics or VLDL concentration. Yet, none of the listed kinetic studies have carefully separated the cases according to severity of diabetes or the degree of obesity.

The average fractional catabolic rate of VLDL triglycerides is normal or slightly decreased in most of the diabetic patient groups described in table I. However, those patients who have marked hypertriglyceridemia show regularly an impaired VLDL clearance [*Nikkilä and Kekki,* 1973; *Kissebah* et al., 1982; *Dunn* et al., 1984]. In accordance with this the lipoprotein lipase activity measured either in postheparin plasma or in adipose tissue is slightly to moderately reduced in type 2 diabetic patients, the change being most prominent in hypertriglyceridemic subjects [*Nikkilä* et al., 1977; *Taylor* et al., 1979; *Taskinen* et al., 1982; *Pfeifer* et al., 1983]. However, in some studies no difference in LPL activity has been found between hypertriglyceridemic diabetics and normolipidemic control subjects [*Abrams* et al., 1982]. The diabetic patients with normal triglyceride levels show also normal lipoprotein lipase activity [*Nikkilä* et al., 1977].

Taken together the above results suggest that the basic abnormality of plasma triglyceride metabolism in type 2 diabetes is overproduction of VLDL particles. This results in elevation of plasma VLDL concentration which may be slight, moderate or severe depending on the removal efficiency and, second, on the excess of VLDL secretion above normal range. Marked hyperlipidemia or fasting chylomicronemia develops only in face of a significant removal defined as a fractional catabolic rate of 30% or less of normal average. The excessive secretion of VLDL into plasma in noninsulin-dependent diabetes is probably accounted for by simultaneous presence of hyperinsulinemia, hyperglycemia and increased release of FFA from mesenteric fat into portal blood. *Howard* et al. [1983] have demonstrated a close positive relationship between VLDL triglyceride synthesis and plasma C-peptide levels.

Effect of Treatment on VLDL Triglyceride Metabolism

The three antidiabetic treatment methods – diet, oral drugs and insulin – are not equally effective in correcting the disturbances of lipoprotein metabolism. Thus, amelioration of hyperglycemia is not necessarily accompanied by normalization of plasma lipoprotein levels. These differences are related to the different mechanisms by which the various forms of treatment decrease blood glucose and they also indicate that hyperglycemia itself is not the crucial factor in the pathogenesis of diabetic lipid disturbances.

Weight reduction is accompanied by improvement of glycemic control and lowering of plasma triglyceride and VLDL concentrations.

Caloric restriction significantly reduces the VLDL triglyceride synthesis without influencing the fractional catabolic rate [*Ginsberg and Grundy*, 1982]. In diabetic lipemia reduction of dietary fat intake is followed by disappearance of chylomicronemia and fall of VLDL levels [*Chait* et al., 1981].

Insulin administration lowers the concentration of VLDL in type 2 diabetics [*Strisower* et al., 1958; *Abrams* et al., 1982; *Dunn* et al., 1984] by decreasing the VLDL production [*Abrams* et al., 1982; *Dunn* et al., 1984] and by stimulating the lipoprotein lipase activity [*Taylor* et al., 1979]. The removal of VLDL is improved by insulin in diabetic patients with lipemia and clearance defect [*Dunn* et al., 1984].

Sulphonylureas are usually little effective against diabetic hypertriglyceridemia [*Nikkilä* et al., 1977; *Breier* et al., 1982; *Walden* et al., 1984]. Biguanides have been shown to decrease VLDL triglyceride production and increase their removal [*Kissebah* et al., 1974] but the effects of various oral antidiabetic drugs have not been adequately compared in larger prospective series.

References

Abrams, J.J.; Ginsberg, H.; Grundy, S.M.: Metabolism of cholesterol and plasma triglycerides in nonketotic diabetes mellitus. Diabetes *31*: 903–910 (1982).

Adlersberg, D.; Wang, C.C.-I.: Syndrome of idiopathic hyperlipemia, mild diabetes mellitus, and severe vascular damage. Diabetes *4:* 210–218 (1955).

Beynen, A.C.; Haagsman, H.P.; Van Golde, L.M.G.; Geelen, M.J.H.: The effects of insulin and glucagon on the release of triacylglycerols by isolated rat hepatocytes are mere reflections of the hormonal effects on the rate of triacylglycerol synthesis. Biochim. biophys. Acta *665:* 1–7 (1981).

Breier, C.; Lisch, H.J.; Sailer, S.: Effect of treatment on the concentration of lipoproteins and the postheparin-lipolytic activity in the plasma of noninsulin-dependent diabetics. Klin. Wschr. *60:* 551–554 (1982).

Chait, A.; Robertson, T.H.; Brunzell, J.D.: Chylomicronemia syndrome in diabetes mellitus. Diabetes Care *4:* 343–348 (1981).

Dunn, F.L.; Raskin, P.; Bilheimer, D.W.; Grundy, S.M.: The effect of diabetic control on very low-density lipoprotein – triglyceride metabolism in patients with type II diabetes mellitus and marked hypertriglyceridemia. Metabolism *33:* 117–123 (1984).

Durrington, P.N.; Newton, R.S.; Weinstein, D.B.; Steinberg, D.: Effects of insulin and glucose on very low density lipoprotein triglyceride secretion by cultured rat hepatocytes. J. clin. Invest. *70:* 63–73 (1982).

Ginsberg, H.; Grundy, S.M.: Very low density lipoprotein metabolism in non-ketotic diabetes mellitus. Effect of dietary restriction. Diabetologia *23:* 421–425 (1982).

Goldberg, A.P.; Chait, A.; Brunzell, J.D.: Postprandial adipose tissue lipoprotein lipase activity in primary hypertriglyceridemia. Metabolism 29: 223–229 (1980).

Greenfield, M.; Kolterman, O.; Olefsky, J.; Reaven, G.M.: Mechanism of hypertriglyceridemia in diabetic patients with fasting hyperglycaemia. Diabetologia 18: 441–446 (1980).

Howard, B.V.; Reitman, J.S.; Vasquez, B.; Zech, L.: Very-low-density lipoprotein triglyceride metabolism in non-insulin-dependent diabetes mellitus. Relationship to plasma insulin and free fatty acids. Diabetes 32: 271–276 (1983).

Howard, B.V.; Savage, P.J.; Bennion, L.J.; Bennett, P.H.: Lipoprotein composition in diabetes mellitus. Atherosclerosis 30: 153–162 (1978).

Kissebah, A.H.; Adams, P.W.; Wynn, V.: Inter-relationship between insulin secretion and plasma free fatty acid and triglyceride transport kinetics in maturity onset diabetes and the effect of phenylethylbiguanide (Phenformin). Diabetologia 10: 119–130 (1974).

Kissebah, A.H.; Alfarsi, S.; Evans, D.J.; Adams, P.W.: Integrated regulation of very low density lipoprotein triglyceride and apolipoprotein-B kinetics in non-insulin-dependent diabetes mellitus. Diabetes 31: 217–225 (1982).

Kolb, F.O.; Lalla, O.F. de; Gofman, J.W.: The hyperlipemias in disorders of carbohydrate metabolism: serial lipoprotein studies in diabetic acidosis with xanthomatosis and in glycogen storage disease. Metabolism 4: 310–317 (1955).

Lewis, B.; Mancini, M.; Mattock, M.; Chait, A.; Fraser, T.R.: Plasma triglyceride and fatty acid metabolism in diabetes mellitus. Eur. J. clin. Invest. 2: 445–453 (1972).

Lopes-Virella, M.F.; Wohltmann, H.J.; Loadholt, C.B.; Buse, M.G.: Plasma lipids and lipoproteins in young insulin-dependent diabetic patients: relationship with control. Diabetologia 21: 216–223 (1981).

Lopes-Virella, M.F.; Wohltmann, H.J.; Mayfield, R.K.; Loadholt, C.B.; Colwell, J.A.: Effect of metabolic control on lipid, lipoprotein, and apolipoprotein levels in 55 insulin-dependent diabetic patients: a longitudinal study. Diabetes 32: 20–25 (1983).

Nikkilä, E.A.: Regulation of hepatic production of plasma triglycerides by glucose and insulin. Proc. VIth Alfred Benzon Symposium, pp. 360–378 (Munksgaard, Copenhagen 1974).

Nikkilä, E.A.: Plasma lipid and lipoprotein abnormalities in diabetes; in Jarrett, Diabetes and heart disease, pp. 133–167 (Elsevier, Amsterdam 1984).

Nikkilä, E.A.; Hormila, P.: Serum lipids and lipoproteins in insulin-treated diabetes. Demonstration of increased high density lipoprotein concentrations. Diabetes 27: 1078–1086 (1978).

Nikkilä, E.A.; Huttunen, J.K.; Ehnholm, C.: Postheparin plasma lipoprotein lipase and hepatic lipase in diabetes mellitus. Relationship to plasma triglyceride metabolism. Diabetes 26: 11–21 (1977).

Nikkilä, E.A.; Kekki, M.: Plasma triglyceride transport kinetics in diabetes mellitus. Metabolism 22: 1–22 (1973).

Patsch, W.; Franz, S.; Schonfeld, G.: Role of insulin in lipoprotein secretion by cultured rat hepatocytes. J. clin. Invest. 71: 1161–1174 (1983).

Pfeifer, M.A.; Brunzell, J.D.; Best, J.D.; Judzewitsch, R.G.; Halter, J.B.; Porte, D., Jr.: The response of plasma triglyceride, cholesterol, and lipoprotein lipase to treatment in non-insulin-dependent diabetic subjects without familial hypertriglyceridemia. Diabetes 32: 525–531 (1983).

Pietri, A.O.; Dunn, F.L.; Grundy, S.M.; Raskin, P.: The effect of continuous subcutaneous insulin infusion on very-low-density lipoprotein triglyceride metabolism in type I diabetes mellitus. Diabetes *32:* 75–81 (1983).

Reaven, G.M.; Mondon, C.E.: Effect of in vivo plasma insulin levels on the relationship between perfusate free fatty acid concentration and triglyceride secretion by perfused rat livers. Hormone metabol. Res. *16:* 230–232 (1984).

Taskinen, M.-R.; Nikkilä, E.A.: Lipoprotein lipase activity of adipose tissue and skeletal muscle in insulin-deficient human diabetes. Relation to high-density and very-low-density lipoproteins and response to treatment. Diabetologia *17:* 351–356 (1979).

Taskinen, M.-R.; Nikkilä, E.A.: Lipoprotein lipase of adipose tissue and skeletal muscle in human obesity: response to glucose and to semistarvation. Metabolism *30:* 810–817 (1981).

Taylor, K.G.; Galton, D.J.; Holdsworth, G.: Insulin-independent diabetes: a defect in the activity of lipoprotein lipase in adipose tissue. Diabetologia *16:* 313–317 (1979).

Topping, D.L.; Mayes, P.A.: The immediate effects of insulin and fructose on the metabolism of the perfused liver. Changes in lipoprotein secretion, fatty acid oxidation and esterification, lipogenesis and carbohydrate metabolism. Biochem. J. *126:* 295–311 (1972).

Vogelberg, K.H.; Gries, F.A.; Moschinski, D.: Hepatic production of VLDL-triglycerides – dependence of portal substrate and insulin concentration. Hormone metabol. Res. *12:* 688–694 (1980).

Walden, C.E.; Knopp, R.H.; Wahl, P.W.; Beach, K.W.; Strandness, E., Jr.: Sex differences in the effect of diabetes mellitus on lipoprotein triglyceride and cholesterol concentrations. New Engl. J. Med. *311:* 953–959 (1984).

Weidman, S.W.; Ragland, J.B.; Fisher, J.N., Jr.; Kitabchi, A.E.; Sabesin, S.M.: Effects of insulin on plasma lipoproteins in diabetic ketoacidosis: evidence for a change in high density lipoprotein composition during treatment. J. Lipid Res. *23:* 171–182 (1982).

Yki-Järvinen, H.; Taskinen, M.-R.; Koivisto, V.A.; Nikkilä, E.A.: Response of adipose tissue lipoprotein lipase activity and serum lipoproteins to acute hyperinsulinemia in man. Diabetologia *27:* 364–369 (1984).

Esko A. Nikkilä, MD, Third Department of Medicine,
University of Helsinki Hospital, SF–00290 Helsinki 29 (Finland)

Glycosylation of Lipoproteins: Chemistry and Biological Implications

R. Flückiger

Department of Research and Internal Medicine, University Clinics, Kantonsspital, Basel, Switzerland

Introduction

All three major lipoprotein fractions contain apoproteins which are glycoproteins: Apo B in the cholesterol-rich low density lipoprotein (LDL), Apo B and Apo C-III in the triglyceride-rich very low density lipoprotein (VLDL), and Apo E which is a constituent of the phospholipid-rich high density lipoprotein (HDL) as well as of LDL and VLDL. In addition to this enzymatic glycosylation, nonenzymatic glycosylation can occur with all apolipoproteins while in circulation.

Recently, the nonenzymatic glycosylation – referred to as glycation in the following – of the apoproteins has received much attention because of its possible pathophysiological relevance to atherosclerosis in diabetes. This is a review of current knowledge concerning enzymatic and nonenzymatic glycosylation of the apolipoproteins with emphasis on the chemistry of these posttranslational modification reactions and their physiological and pathophysiological role.

Enzymatic Glycosylation

Apoprotein B: N-Glycosidically Linked Oligosaccharide

The characteristic protein moiety of human LDL, Apo B, contains 5–9% carbohydrate consisting of galactose, mannose and sialic acid. Two glycopeptides were isolated and their monosaccharide sequence determined by sequential enzymatic digestion, periodate oxidation and partial acid hydrolysis [32]. The proposed structures of the two glycopeptides representing at least 50% of the carbohydrate of LDL is the following:

$$\text{Man} \xrightarrow[(1 \to 2)]{\alpha} (\text{Man})_3 \xrightarrow{\alpha} \text{Man} \xrightarrow{\beta} (\text{Man, GluNAc, GluNAc}) \longrightarrow \text{Asn}$$

I

$$\text{NANA} \longrightarrow \text{Gal} \xrightarrow{\beta} \text{GluNAc} \xrightarrow{\beta} \text{Man}$$
$$\searrow \alpha$$
$$(\text{Man-Man, GluNAc}) \longrightarrow \text{Asn}$$
$$\nearrow \alpha \uparrow \alpha (1 \to 2)$$
$$\text{NANA} \longrightarrow \text{Gal} \xrightarrow{\beta} \text{GluNAc} \xrightarrow{\beta} \text{Man}$$
$$\text{Man}$$

II

It is not clear whether additional oligosaccharide chains occur in LDL and how many of the above chains are present on an LDL molecule. Also it is not known whether different subpopulations of LDL each with different types of oligosaccharide chains exist. Available data indicate that the branched oligosaccharide type II chain predominates over the linear type I chains. Such N-linked oligosaccharide chains are synthesized by membrane bound glycosyltransferase of the rough endoplasmic reticulum with dolichol serving as membrane anchor for the growing oligosaccharide chain. The oligosaccharide precursor is transferred to the nascent polypeptide as it is extruded into the lumen of the endoplasmic reticulum. The asparagine is usually in the sequence Asn-X-Thr/Ser. After the transfer to the acceptor protein, the oligosaccharide is processed by several glycosidases during the transport of the glycoprotein to the Golgi apparatus [for review, see 5].

The biological function of the oligosaccharide chain of circulating glycoproteins is thought to lie in biological recognition. Removal of the terminal sialic acid residues reduces the half-life of most of these glycoproteins because the penultimate galactose residue is recognized by the galactose receptor on the hepatocytes allowing internalization and degradation [2].

The validity of this concept has *not* been unequivocally demonstrated for Apo B. In vivo measurements of turnover rates of desialylated and control lipoproteins in rat [31], rabbit [4], and pig [3] have yielded contradictory results. Kinetic data for injected human desialylated LDL in human subjects show that the first part of a biphasic disappearance curve shows a more rapid disappearance, with a clearance rate which is 52% higher for desialylated LDL than that of the control [21]. Thus, binding of desialylated human LDL may be enhanced but not its catabolism. Uptake of human LDL by cultured human smooth muscle cells was found to be much faster after sialic acid removal [11]. Enzymatic removal of virtually all carbohydrate residues from human LDL did not alter their binding to cultured human fibroblasts [29]. Also, catabolism of neuraminidase-treated human LDL by rat heptatocytes was not altered [11]. Another biological function for the oligosaccharide chains could be in cellular handling of the protein up to the point of secretion. This role is suggested since the liver of orotic acid fed rats is unable to secrete LDL or VLDL [22, 36] probably because this substance prevents adequate glycosylation. In these animals the nonglycosylated apolipoproteins accumulate in liposomes with the exception of one of a pair of apolipoproteins, possibly a protein which is normally not glycosylated. The reason for this is an inhibition of glycosylation probably because of the increased UDP-N-acetylglucosamine concentration. Dietary adenine prevents the appearance of a fatty liver [23] and also the rise in UDP-N-acetylglucosamine.

Apoprotein C-III: O-Glycosidically Linked Oligosaccharide

In Apo C-III a single O-glycosidically linked carbohydrate moiety is present. It contains 1 mol each of galactosamine and galactose and 0–2 mol sialic acid/mol protein. The oligosaccharide chain is attached to threonine-74 close to the carboxy terminus [6]. Its sequence and the types of linkages have not been determined.

The three polymorphic forms of Apo C-III which occur physiologically arise because of the different sialic acid content of the respective Apo C-III species. The relative concentration of the different Apo C-III forms has been determined in healthy subjects [1]. The most abundant form containing 1 sialic acid represents some 50% of the total Apo C-III and the sialic acid free form 25%. The synthesis of O-linked oligosaccharides occurs by sequential glycosylation during the transport of the protein from the endoplasmic reticulum to the Golgi apparatus. The types and activities of the glycosyltransferases determine the structure of the oligosaccharide chain. A specific polypeptide sequence around the threonine or serine does not appear to be required for glycosylation and incomplete glycosylation is often observed. Synthesis begins with the transfer of N-acetylgalactosamine to the hydroxyl group of serine or threonine. In keeping with this mechanism, the linkage in Apo C-III is probably to N-acetylgalactosamine. N-acetylgalactosamine was also identified in the Lp(a) lipoprotein, a lipoprotein variant. The protein moiety of Lp(a) consists of about 65% Apo B, 15% albumin, and 20% of the apoprotein Lp(a). Apo Lp(a) contains 2.5 times more carbohydrate than Apo LDL. It contains about 6 times more sialic acid than Apo LDL, 3 times more hexosamines and twice the amount of hexose [10]. While occurring in the LDL density range, Lp(a) has the same pre-beta electrophoretic mobility as VLDL, probably due to its high content of sialic acid. The presence of N-acetylgalactosamine indicates that carbohydrate may be bound to serine or threonine.

Apoprotein E: Glycosidic Linkage Not Determined

The 'arginine-rich' apoprotein first identified in VLDL, Apo E, occurs in all lipoprotein fractions. Its carbohydrate content ranges from 3 to 5.4% and it contains sialic acid, mannose, galactose, N-acetylglucosamine and N-acetylgalactosamine. Its sialic acid content ranges from 0.79 to 1.69 mol/mol of Apo E in different preparations [16]. This variable sialic acid content is the cause of part of the microheterogeneity of Apo E. Additional heterogeneity results from a genetic polymorphism [40]. Desialylation converts some of the more acidic isoforms resolved by high resolution isoelectric focusing to the three major isoforms of Apo E [15, 40]. The physiological role of the oligosaccharide chain in Apo E is unknown at present. With regard to a pathophysiological role it is interesting to note that a cholesterol-rich diet sharply increased Apo E levels in guinea pigs concurrent with an increase of Apo E isoforms differing in sialic acid content [15].

Also, in type II diabetic patients a slight shift to a higher proportion of Apo E-I and Apo E-II which are enriched in sialo-Apo E [41] has been reported. Based on this evidence a defect in the desialylation of Apo E in diabetic patients was assumed to exist [35].

Nonenzymatic Glycosylation (Glycation) of Lipoproteins

The occurrence of protein glycosylation was recognized as a physiological process when the structure and synthesis of one of the minor hemoglobin components, hemoglobin A1c, which is present in increased concentration in diabetics was clarified [for review, see 7]. In this reaction the open chain structure of glucose and other reactive carbonyl compounds such as other aldoses and ketoses [8, 24] react with amino groups to initially form a Schiff's base adduct or aldimine (II) which can rearrange to the corresponding ketoamine (III):

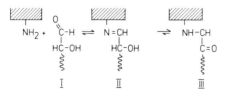

This rearrangement (Amadori) is slow and only longer lived proteins such as hemoglobin which stays in the circulation for 120 days get glycated to an appreciable extent.

Of particular interest for the experiments with lipoprotein fractions which were obtained by in vitro glycation is the observation that the sites of glycation differ depending on whether glycation occurred in vivo or in vitro [28]. A concept which is emerging from available data indicates that Schiff's base formation primarily depends on the pK of the amino group and that ketoamine formation depends from the proximity of a catalytic proton donor/acceptor group. Based on this rationale all sites of glycation in vivo can be predicted and the extent to which glycation occurs at these sites can be estimated. It also explains why in albumin the principal site of glycation is lysine-525 and not lysine-199, the amino group with a particularly low pK [12]. Lysine-525 in albumin occurs in a lysine-lysine sequence, as do the second most reactive lysines in the alpha and beta chain of hemoglobin [28]. The reactive Lys-Lys sequence is also present in Apo A-I and Apo A-II together with the other potentially reactive sequence Lys-X-Lys. Lys-X-Lys also occurs in Apo C [for sequence data, cf. 9]. One would therefore expect the apoproteins to be good candidates for glycation. Interestingly, the LDL receptor binding region of Apo E, His(140)-Leu-Arg-Lys-Leu-Arg-Lys-Arg-Leu-Leu-Arg(150), would be predicted to be readily glycatable. Apart from causing site specificity, the proximity of a catalytic center also seems to influence the rate of the reaction. The overall rate of glycation of albumin is approximately 10 times greater than that of hemoglobin [12].

Because of the short residence time in circulation of the apolipoproteins – the half-life in man for HDL is 3–6 days [26] and the mean transit time for LDL is 2.5 days [21] – their degree of glycation is quite small. The figures obtained differ somewhat depending on the methodology used. Estimates for the extent of glycation of LDL range from 0.5 to 4 mol glucose/mol protein in normals and up to 4 times this amount in diabetics [27, 37]. Similarly low figures were reported for HDL of normoglycemic persons with values ranging from 1.1 to 3.8% glycosylated lysine residues [37]. The extent of glycation of LDL of 0.5 mol glucose/mol protein agrees well with that predicted from the kinetics of in vitro glycosylation while glycation of HDL would be overestimated by a factor of 10–20 [33].

Influence of Glycation on Biological Properties of HDL and LDL

The glycation-induced changes in biological properties of HDL and LDL are quite small and as a consequence most experiments were performed with heavily glycated lipoprotein fractions. These are obtained by incubating the lipoprotein with excess glucose in the presence of the reducing agent cyanoborohydride for prolonged periods of time. The cyanoborohydride 'traps' the Schiff's base formed in the first, rapid step of the glycation reaction. 70% of lysine residues in LDL and 60% in HDL could be modified in this manner [37, 38].

HDL: In guinea pigs the rate of clearance of heavily glycated HDL is increased. The mechanism accounting for the enhanced catabolism of heavily glycated HDL is unknown. Its uptake by macrophages is unaltered nor is uptake or degradation by cultured hepatocytes changed [37]. Interestingly, glycation of as few as 2% lysine residues in HDL was sufficient to increase its fractional catabolic rate by about 8% in guinea pigs. Whether this glycation-induced effect contributes significantly to the reduced HDL levels in diabetics remains to be established.

LDL: Cellular cholesterol metabolism is regulated by a specific LDL receptor pathway. Endocytosis of the cholesterol-rich LDL by this pathway regulates both the rate of accumulation and the rate of removal of tissue cholesterol deposits and hence may influence the development of atherosclerosis [13]. The binding of lipoproteins to their receptors is determined by the protein moieties and can be changed by selective modification of lysine or arginine residues of the apoproteins [19, 20]. Chemical modification prevents binding, internalization and degradation by high affinity receptors on human skin fibroblasts.

In vitro glycation to an unphysiologically high extent of LDL also results in a decreased binding, uptake and degradation of the LDL by the high affinity process in normal human fibroblasts and a decreased clearance in vivo [14, 24, 25, 30, 38]. In guinea pigs the fractional catabolic rate and the degree of LDL glycation is inversely related. Extrapolation from these data suggest that complete inhibition of receptor-mediated LDL uptake occurs if 22% of all lysines are glycated. LDL glycation, to an extent comparable to that occurring in diabetics (2–5% glycosylated lysine residues), decreased clearance in guinea pigs by 5–25% [30], and also in man [39]. This is an indication that glycation of LDL in diabetics may indeed cause a decreased clearance. However, turnover data with moderately glycated LDL in human diabetics are needed to establish such a role.

The concept of an altered metabolism due to LDL glycation is supported by the finding that the fractional catabolic rate of LDL Apo B is reduced by approximately 20% in moderately severe diabetics [17]. However, the increase of the catabolic rate in mild diabetes is difficult to explain with the 'glycation hypothesis'. Other conflicting evidence was obtained when the metabolism by fibroblasts of LDL isolated from diabetics in poor glycemic control was compared to that isolated after achieving good metabolic control [18]. The degree of glycemic control not only influenced LDL metabolism but also LDL composition. A significant decrease of the triglyceride/protein ratio in postcontrol LDL was apparent. The main difference in fibroblast interaction and processing concerned intracellular degradation while binding of LDL to fibroblasts was not affected [18]. Available data therefore appear insufficient to demonstrate a pathophysiological role of lipoprotein glycosylation and glycation in accelerated atherosclerosis in diabetes.

Research efforts focusing on *all* aspects of the lipoprotein fractions – synthesis and degradation, lipid and protein composition, enzymatic and nonenzymatic glycosylation, as well as the interaction with tissue components – are needed to establish the mechanism of atherogenesis in diabetes on a molecular basis.

References

1 Albers, J.J.; Scanu, A.M.: Isoelectric fractionation and characterization of polypeptides from human serum very low density lipoproteins. Biochim. biophys. Acta 236: 29–37 (1971).

2 Ashwell, G.; Morell, A.G.: The role of surface carbohydrates in the hepatic recognition and transport of circulating glycoproteins. Adv. Enzymol. *41:* 99–128 (1974).
3 Attie, A.D.; Weinstein, D.B.; Freeze, H.H.; Pittman, R.C.; Steinberg, D.: Unaltered catabolism of desialylated low-density lipoprotein in the pig and in cultured rat hepatocytes. Biochem. J. *180:* 647–654 (1979).
4 Avila, E.M.; Lopez, F.; Camejo, G.: Properties of low density lipoproteins related to its interaction with arterial wall components – in vitro and in vivo studies. Artery *4:* 36–60 (1978).
5 Berger, E.G.; Buddecke, E.; Kamerling, J.P.; Kobata, A.; Paulson, J.C.; Vliegenthart, J.F.G.: Structure, biosynthesis and functions of glycoprotein glycans. Experientia *38:* 1129–1258 (1982).
6 Brewer, H.B.; Shulman, R.; Herbert, P.; Ronan, R.; Wehrly, K.: The complete amino acid sequence of alanine apolipoprotein (apoC-III), an apolipoprotein from human plasma very low density lipoproteins. J. biol. Chem. *249:* 4975–4984 (1974).
7 Bunn, H.F.; Gabbay, K.H.; Gallop, P.M.: The glycosylation of hemoglobin: relevance to diabetes mellitus. Science *200:* 21–27 (1978).
8 Bunn, H.F.; Higgins, P.J.: Reaction of monosaccharides with proteins: possible evolutionary significance. Science *213:* 222–224 (1981).
9 Dayhoff, M.O.: Atlas of protein sequence and structure, vol. 5 (National Biomedical Research Foundation, Silver Spring 1972).
10 Ehnholm, C.; Garoff, H.; Renkonen, O.; Simons, K.: Protein and carbohydrate composition of Lp(a) lipoprotein from human plasma. Biochemistry *11:* 3229–3232 (1972).
11 Filipovic, I.; Schwarzmann, G.; Mraz, W.; Weigandt, H.; Buddecke, E.: Sialic-acid content of low density lipoproteins controls their binding and uptake by cultured cells. Eur. J. Biochem. *93:* 51–55 (1979).
12 Garlick, R.L.; Mazer, J.S.: The principal site of nonenzymatic glycosylation of human serum albumin in vivo. J. biol. Chem. *258:* 6142–6146 (1983).
13 Goldstein, J.L.; Brown, M.S.: The low-density lipoprotein pathway and its relation to atherosclerosis. A. Rev. Biochem. *46:* 897–930 (1979).
14 Gonen, B.; Baenziger, J.; Schonfeld, G.; Jacobson, D.; Farrar, P.: Nonenzymatic glycosylation of low density lipoproteins in vitro. Effects on cell-interactive properties. Diabetes *30:* 875–878 (1981).
15 Guo, L.S.S.; Hamilton, R.L.; Kane, J.P.; Fielding, C.J.; Chi Chen, G.: Characterization and quantitation of apolipoproteins A-I and E of normal and cholesterol-fed guinea pigs. J. Lipid Res. *23:* 531–542 (1982).
16 Jane, R.S.; Quartfordt, S.H.: The carbohydrate content of apolipoprotein E from human very low density lipoproteins. Life Sci. *25:* 1315–1324 (1979).
17 Kissebah, A.H.; Alfarsi, S.; Evans, D.J.; Adams, P.W.: Plasma low density lipoprotein transport kinetics in noninsulin-dependent diabetes mellitus. J. clin. Invest. *71:* 655–667 (1983).
18 Lopes-Virella, M.F.; Sherer, G.K.; Lees, A.M.; Wohltmann, H.; Mayfield, R.; Sagel, J.; LeRoy, E.C.; Colwell, J.A.: Surface binding, internalization and degradation by cultured human fibroblasts of low density lipoproteins isolated from type 1 (insulin-dependent) diabetic patients: changes with metabolic control. Diabetologia *22:* 430–436 (1982).

19 Mahley, R.W.; Innerarity, T.L.; Pitas, R.E.; Weisgraber, K.H.; Brown, J.H.; Gross, E.: Inhibition of lipoprotein binding to cell surface receptors of fibroblasts following selective modification of arginyl residues in arginine-rich and B apoproteins. J. biol. Chem. 252: 7279–7287 (1977).

20 Mahley, R.W.; Innerarity, T.L.; Weisgraber, K.H.; Oh, S.Y.: Altered metabolism (in vivo and in vitro) of plasma lipoproteins after selective chemical modification of lysine residues of the apoproteins. J. clin. Invest. 64: 743–750 (1979).

21 Malmendier, C.L.; Delcroix, C.; Fontaine, M.: Effect of sialic acid removal on human low density lipoprotein catabolism in vivo. Atherosclerosis 37: 277–284 (1980).

22 Pottenger, L.A.; Getz, G.S.: Serum lipoprotein accumulation in the livers of orotic acid-fed rats. J. Lipid Res. 12: 450–459 (1971).

23 Pottenger, L.A.; Frazier, L.E.; DuBien, L.H.; Getz, G.S.; Wissler, R.W.: Carbohydrate composition of lipoprotein apoproteins isolated from rat plasma and from the livers of rats fed orotic acid. Biochem. biophys. Res. Commun. 54: 770–776 (1973).

24 Sasaki, J.; Arora, V.; Cottam, G.L.: Nonenzymatic galactosylation of human LDL decreases its metabolism by human skin fibroblasts. Biochem. biophys. Res. Commun. 108: 791–796 (1982).

25 Sasaki, J.; Cottam, G.L.: Glycosylation of human LDL and its metabolism in human skin fibroblasts. Biochem. biophys. Res. Commun. 104: 977–983 (1982).

26 Schaefer, E.J.; Levy, R.J.: Composition and metabolism of high-density lipoproteins. Prog. biochem. Pharmacol., vol. 25, pp. 200–215 (Karger, Basel 1979).

27 Schleicher, E.; Deufel, T.; Wieland, O.H.: Non-enzymatic glycosylation of human serum lipoproteins – elevated epsilon-lysine glycosylated low density lipoprotein in diabetic patients. FEBS Lett. 129: 1–4 (1981).

28 Shapiro, R.; McManus, M.J.; Zalut, C.; Bunn, H.F.: Sites of nonenzymatic glycosylation of human hemoglobin A. J. biol. Chem. 255: 3120–3127 (1980).

29 Shireman, R.B.; Fisher, W.R.: The absence of a role for the carbohydrate moiety in the binding of apolipoprotein B to the low density lipoprotein receptor. Biochim. biophys. Acta 572: 537–540 (1979).

30 Steinbrecher, U.P.; Witztum, J.L.: Glucosylation of low-density lipoproteins to an extent comparable to that seen in diabetes slows their catabolism. Diabetes 33: 130–134 (1984).

31 Swaminathan, N.; Aladjem, F.: The carbohydrate moiety of human serum low density lipoproteins. Fed. Proc. 33: 1585 (1974).

32 Swaminathan, N.; Aladjem, F.: The monosaccharide composition and sequence of the carbohydrate moiety of human serum low density lipoproteins. Biochemistry 15: 1516–1522 (1976).

33 Vaughan, L.; Fischer, R.W.; Zimmermann, D.R.; Winterhalter, K.H.: Nonenzymatic glucosylation of proteins: a new and rapid solution for in vitro investigation. FEBS Lett. 173: 173–178 (1984).

34 Weisgraber, K.H.; Innerarity, T.L.; Mahley, R.W.: Role of the lysine residues of plasma lipoproteins in high affinity binding to cell surface receptors on human fibroblasts. J. biol. Chem. 253: 9053–9062 (1978).

35 Weisweiler, P.; Jüngst, D.; Schwandt, P.: Quantitation of apolipoprotein E isoforms in diabetes mellitus. Hormone metabol. Res. 15: 201 (1983).

36 Windmueller, H.G.; Levy, R.I.: Total inhibition of hepatic beta-lipoprotein production in the rat by orotic acid. J. biol. Chem. *242:* 2246–2254 (1967).
37 Witztum, J.L.; Fisher, M.; Pietro, T.; Steinbrecher, U.P.; Elam, R.L.: Nonenzymatic glucosylation of high-density lipoprotein accelerates its catabolism in guinea pigs. Diabetes *31:* 1029–1032 (1982).
38 Witztum, J.L.; Mahoney, E.M.; Branks, M.J.; Fisher, M.; Elam, R.; Steinberg, D.: Nonenzymatic glucosylation of low-density lipoprotein alters its biologic activity. Diabetes *31:* 283–291 (1982).
39 Witztum, J.L.; Steinbrecher, U.P.; Fisher, M.; Kesaniemi, A.: Nonenzymatic glucosylation of homologous low density lipoprotein and albumin renders them immunogenic in the guinea pig. Proc. natn. Acad. Sci. USA *80:* 2757–2761 (1983).
40 Zannis, V.J.; Breslow, J.L.: Human very low density lipoprotein apolipoprotein E isoprotein polymorphism is explained by genetic variation and posttranslational modification. Biochemistry *20:* 1033–1041 (1981).
41 Zannis, V.I.; Breslow, J.L.: Apolipoprotein E. Mol. cell. Biochem. *42:* 3–20 (1982).

R. Flückiger, PhD, Zentrum für Lehre und Forschung,
Diabetologie, Hebelstrasse 20, CH-4031 Basel (Switzerland)

Pathophysiology of Low Density Lipoprotein and High Density Lipoprotein Glucosylation

Y. Antero Kesäniemi

Second Department of Medicine, University of Helsinki, Helsinki, Finland

Several abnormalities in lipoprotein metabolism have been reported in patients with diabetes mellitus particularly in type II diabetics such as hypertriglyceridemia, hypercholesterolemia, increased very low density lipoprotein (VLDL) triglyceride (TG) and apolipoprotein B synthesis, enhanced cholesterol and bile acid production, diminished high density lipoprotein (HDL) cholesterol concentration and lowered lipoprotein lipase activity [1–5]. Since many plasma proteins undergo posttranslational nonenzymatic glucosylation in hyperglycemia, lipoproteins might also be exposed to this modification and some of the abnormalities in lipid and lipoprotein metabolism in diabetes could be related to the nonenzymatic glucosylation of lipoproteins. Thus, lipoprotein glucosylation might play a role in the development of atherosclerosis known accelerated in diabetics [6] in the same way as glucosylation of several other proteins has been suggested to contribute to a number of pathophysiologic complications of diabetes such as cataract formation, neuropathy and connective tissue abnormalities [7].

Reaction for Nonenzymatic Glucosylation of LDL and HDL

Glucose forms a labile intermediate Schiff base with the epsilon amino group of lysine residues and the Schiff base undergoes an Amadori rearrangement to form the more stable ketoamine and hemiketal adducts [8, 9]. The Schiff base, ketoamine and hemiketal products can all exist in vivo. When glucosylation of LDL and HDL is carried out in vitro in the presence of a reducing agent such as $NaCNBH_3$, the labile Schiff base is immediately and quantitatively reduced to glucitollysine [8, 10]. If glucosylation of lipoproteins is performed in vitro in the absence of a reducing agent, the labile Schiff base undergoes an Amadori rearrangement to form the stable ketoamine and hemiketal adducts as in vivo [10]. Only the Schiff is reduced to glucitollysine by NaCNBH$_3$ reduction, whereas each of the other products, Schiff base, ketoamine and hemiketal can be reduced to glucitollysine by NaBH$_4$ [10]. The hemiketal reduced by NaBH$_4$ can also result in the generation of a mannositollysine residue, the epimer of a glucitollysine residue [10].

Metabolic Consequences of LDL Glucosylated Heavily in vitro

In vitro glucosylation of LDL with a reducing agent results in irreversible glucosylation of 6–60% of lysine residues depending on the time of incubation and the concentration of glucose [11–13]. Heavily glucosylated LDL (45–60% of lysine residues) has greater mobility on agarose electrophoresis than normal LDL [11]. Interestingly, LDL isolated from some diabetic subjects also has similar enhanced mobility [11]. Metabolism of glucosylated LDL differs from that of control LDL in several ways. It is not taken up and degraded by the high affinity LDL receptor pathway in normal skin fibroblasts [11–15]. The uptake and degradation of glucosylated LDL is not accompanied by a reduction in hydroxymethylglutaryl-coenzyme A reductase activity [11] or an increase in acyl coenzyme A:cholesterol acyltransferase activity (ACAT) [11, 14]. In addition, degradation of glucosylated LDL by mouse peritoneal macrophages is not faster than that of native LDL [11]. Thus, heavy glucosylation of LDL results in a blockade of LDL degradation by the LDL receptor pathway but does not stimulate the LDL clearance by the nonreceptor mechanisms.

Clearance of LDL from plasma is mediated via a high-affinity LDL receptor-dependent pathway as well as one or more LDL receptor-independent pathways [16, 17]. To estimate the extent of LDL receptor-dependent and LDL receptor-independent pathway(s) in vivo, *Mahley* et al. [18] and *Shepherd* et al. [19] introduced the concept of performing a double-labeled turnover study using native LDL as a tracer of total LDL catabolism and a chemically modified LDL as a tracer of receptor-independent catabolism. Since nonenzymatic glucosylation of LDL in the presence of NaCNBH$_3$ irreversibly blocks the lysine residues of LDL, glucosylated LDL is not degraded by the LDL receptor of fibroblasts and its degradation by macrophages is similar to that of native LDL, glucosylated LDL should be a good tracer of LDL receptor-independent catabolism and if combined with a tracer of total LDL catabolism, should enable one to calculate the extent of LDL receptor-dependent catabolism. Animal studies showed that the clearance of glucosylated LDL is identical to the clearance of methyl LDL [20], LDL where lysine residues are blocked by methyl groups. Also, the turnover of glucosylated LDL is similar to that of native LDL in the Watanabe heritable hyperlipidemic rabbit, an animal model for homozygous familial hypercholesterolemia (HFH), which has <5% of normal LDL receptors [20]. Finally, our several studies in various human subjects have shown that the fractional catabolic rate of glucosylated LDL ranging from 0.071 to 0.168 day^{-1} in these subjects is exactly in the range observed for LDL turnover of native LDL in patients with HFH (range of 0.06–0.178 day^{-1} [21, 22]. Similar findings have been reported by *Bilheimer* et al. [23] using glucosylated LDL in patients with heterozygous familial hypercholesterolemia. These studies have also indicated that quantitatively 66–80% of LDL is degraded by the receptor-dependent pathway in normal man [21, 22]. Overall, these data suggest that glucosylated LDL serves as a good tracer for the quantitation of the receptor-independent catabolism of LDL in animals and in man. Other kind of data has recently been reported by the workers [24] who clearly modified the methodology that we originally described for the glucose derivatization of LDL [21].

Nonextensive Glucosylation of LDL

Glucosylation of LDL lysine residues is obviously not as extensive in vivo as described above and the products of this modification are different from those of the in vitro glucosylation using a reducing agent. Therefore, we can ask if glucosylation of LDL to an extent comparable to that seen in diabetes mellitus has any effect on LDL metabolism. The presence of glucosylated LDL in the plasma of both euglycemic and diabetic subjects has been demonstrated [10, 11, 13, 25]. Estimations of over 1.0% of lysine residues being glucosylated in normal LDL and 2- to 10-fold more in diabetic LDL have been reported [10, 11, 25]. *Steinbrecher and Witztum* [26] showed that the degradation of glucosylated LDL in cultured fibroblasts is inhibited in proportion to the estimated degree of glucosylation. The same workers also studied turnover in guinea pigs of LDLs with various amounts of lysine residues glucosylated and found that modification of as few as 2–5% of lysines decreased LDL catabolism by 5–25%, and the degree of inhibition of catabolism was linearly related to the extent of LDL glucosylation [26]. These workers emphasize that all the glucosylated LDL preparations with less than 5% glucosylation were prepared by incubation of LDL and glucose without $NaCNBH_3$ indicating that these glucosylated LDL products should be equal to those seen in diabetic plasma. Overall, these results suggest that the extent of LDL glucosylation that can occur in diabetes may slow LDL catabolism and thereby enhance plasma LDL concentration.

LDL Catabolism in Diabetic Patients

Since even minor glucosylation (2–5%) of LDL lysine residues seems to have an inhibitory effect on LDL catabolism in animal studies [26] we might ask if LDL catabolism in diabetics is different from that in euglycemic subjects. *Kissebah* et al. [27] studied a group of patients with noninsulin-dependent diabetes with various degree of obesity, insulin resistance and hyperlipidemia. These workers found that the patients with mild diabetes cleared their LDL at a faster rate (FCR = 0.67 ± 0.04 day^{-1}) than normal subjects (FCR = 0.42 ± 0.02 day^{-1}) whereas the patients with moderately severe diabetes had a significant reduction in LDL apoprotein B FCR (0.31 ± 0.02 day^{-1}) and higher plasma LDL levels. The same differences seemed to be true whether the patients in each group were obese or nonobese. These findings may be related to several factors, one of them being nonenzymatic glucosylation of LDL. It is not known how extensively LDL or other proteins like hemoglobin were glucosylated in the diabetic patients of the study by *Kissebah* et al. [27] but presumably the subjects with moderately severe diabetes having a mean fasting plasma glucose value of 181 mg/dl had more glucosylated proteins than the patients with mild diabetes and an average fasting glucose of 127 mg/dl. Obviously other factors may also be important for the differences in the LDL catabolic rates. One of these could be plasma insulin level and these workers actually found a significant correlation (r = 0.89) between plasma insulin response and the FCR value for LDL ApoB among all the diabetic subjects [27]. Insulin is known to enhance receptor-mediated degradation of LDL in cultured fibroblasts [28] and insulin infusion is associated with increased in vivo catabolism of LDL [29].

Patients with insulin-dependent diabetes often have unsatisfactory glycemic control and correspondingly large amounts of glucosylated plasma proteins as indicated by high HbA_1 levels. Glucosylation of LDL might also be remarkable in these individuals. We recruited 8 insulin-dependent diabetics and 6 controls for LDL turnover study. The diabetics were of the same age (34 ± 4 (SE) years), sex (4 females, 4 males) and weight (71 ± 4 kg) as the controls (age = 47 ± 6; sex = 3 F, 3 M; weight = 67 ± 8). HbA, ranged between 7.7 and 12.7% among the diabetics and most of them had plasma C-peptide levels less than 0.15 µg/l. The mean plasma LDL cholesterol concentrations were equal in the diabetics (2.8 ± 0.2 mmol/l) and controls (2.7 ± 0.3). The VLDL triglyceride levels were also similar in the diabetic patients (0.44 ± 0.07 mmol/l) and control subjects (0.43 ± 0.08). The mean LDL FCR values for the diabetic patients (0.410 ± 0.017 day^{-1}) were also very close to the results for the control subjects (0.440 ± 0.014 day^{-1}) indicating that the clearance rate of LDL was not significantly lowered in insulin-dependent diabetics. Furthermore, the FCR values for LDL were not correlated with the HbA_1 levels among the diabetic patients. Thus, the insulin-dependent diabetics seem to be able to catabolize their LDL at quite a normal rate, and actually these patients may often have low normal LDL levels particularly in good glycemic control [30]. These findings may be related to the high peripheral hyperinsulinemia during insulin treatment and to the stimulatory effect of insulin on the LDL receptor activity [28, 29]. It is therefore possible that the net effect of two simultaneous factors, one being glucosylation of 2–10% of LDL lysine residues and the other insulin-induced stimulation of the LDL receptor activity, may still result in quite a normal LDL clearance from the plasma. Of course, other factors like the production rate of VLDL and LDL and the conversion of VLDL to IDL and LDL also regulate plasma LDL levels.

Clearance of Heavily Glucosylated LDL in Diabetic Subjects

LDL metabolism can obviously be modified by a number of factors in diabetes. Diabetics might also degrade more of their LDL by the receptor-independent pathways than euglycemic subjects. Therefore, we decided to quantitate the receptor-mediated and receptor-independent catabolism of LDL in diabetics. This was carried out by a simultaneous injection of control LDL and LDL glucosylated heavily with NaCHBH$_3$. When control LDL and glucosylated LDL were injected into euglycemic subjects the clearance of the glucosylated LDL was very slow with an average FCR value of 0.11 day^{-1} and only 20% of the rate for control LDL [31]. When glucosylated LDL was injected into 4 subjects with adult-onset diabetes, a similar slow clearance of the glucosylated LDL tracer was noted in 1 of them with the FCR value of 0.09 day^{-1} versus an FCR for control LDL of 0.59 day^{-1}. In the 3 other diabetic patients the glucosylated LDL was cleared at a slower rate than the control LDL during the first 4–8 days but thereafter the glucosylated tracer rapidly disappeared from the plasma [31]. Furthermore, even the initial clearance rate of glucosylated LDL in these 3 diabetics was more rapid than in the subjects not showing the fast disappearance of the glucosylated tracer. This pattern of clearance suggested an immune-mediated mechanism for LDL removal, and the receptor and nonreceptor pathways for LDL catabolism could not be quantitated in these diabetic individuals.

Nonenzymatic glucosylation of homologous LDL and albumin renders these proteins immunogenic in the guinea pigs and leads to rapid immune-mediated clearance of glucosylated LDL [32]. Using solid-phase radioimmunoassay (RIA) autoantibodies against glucosylated LDL could also be demonstrated in the preinjection plasma of the diabetics studied particularly in those showing the rapid clearance of the glucosylated tracer [31]. Antibody titers were low, but specificity determinations showed that almost 80% of the amount bound in the absence of competitor could be displaced by the glucosylated LDL tracer itself but also by glucosylated, reduced HDL and even by glucitollysine. Thus, these antibodies cross-react with other glucosylated lysine epitopes and recognize glucitollysine as well as glucitollysine adducts of many proteins. Antibody class varied among the diabetic patients but in all of them at least one class of immunoglubulins, IgG, IgA or IgM, could be demonstrated with specificity to glucosylated, reduced LDL. It is important to note that these autoantibodies could be found in the plasma before the injection of the glucosylated tracer. This suggests that glucitollysine adducts of LDL or other proteins must have existed in these patients, and that they were immunogenic. However, a reductase capable of generating such glucitollysine adducts in vivo has not yet been described. At any rate, *Curtiss and Witztum* [10], using monoclonal antibodies specific for glucitollysine, observed small amounts of immunoreactivity in untreated plasmas of diabetic subjects. It is also possible, however, that the open-chain ketoamine form was the actual immunogen but that the elicited antibodies cross-reacted with the open-chain glucitol adduct, which is present in large amounts in glucosylated, reduced LDL but only in limited quantity in glucosylated, nonreduced LDL [31].

The finding of autoantibodies against glucosylated proteins in the plasma of diabetics suggests that this posttranslational modification is recognized as foreign by the immune system of at last some individuals. The prevalence and specificity of such antibodies directed against glucosylated proteins in diabetic subjects is not known at the moment. It is noteworthy, however, that the antibodies are directed specifically against glucitollysine, a remarkably small epitope, and are thus capable of reacting with many other glucosylated proteins. Interestingly, *Bassiouny* et al. [33] have recently shown that immunization of rats with reductively glucosylated homologous collagen also results in the formation of antibodies specific for glucitollysine. The pathophysiologic consequences of these antibodies might include formation of circulating immune complexes with modified plasma proteins resulting, for example, in vascular injury. However, if these antibodies also recognize glucose-modified tissue proteins like glucosylated collagen, basement membrane structures and myelin known to exist in diabetes and aged people [7] they may have an important role in the development of several complications like damage of glomerular function, loss of axonal transport and stiffening of the arterial walls, etc.

Glucosylated HDL

HDL also has a plasma residence time long enough to allow posttranslational nonenzymatic glucosylation. Therefore, it is of interest to elucidate the metabolic consequences of glucosylated HDL. *Witztum* et al. [34] glucosylated HDL in vitro together with a reducing agent and noticed that glucosylated HDL had enhanced electrophoretic mobility on agarose. These workers reported that most of the incorporated glucose was localized to apoprotein A-I, but other apoproteins were also glucosylated. In contrast to glucosylated LDL, with increasing amounts of glucose incorporated into HDL there was a proportionate increase in the rate of clearance of glucosylated HDL when injected into guinea pigs. When as few as 2% of lysines were glucosylated there was still an 8% increase in the rate of HDL clearance, but with 60% of lysines derivatized clearance of glucosylated HDL was 60% faster than that of control HDL [34]. Interestingly, these investigators did not find an increased uptake of glucosylated HDL by macrophages. At any rate, they suggested that increased clearance of glucosylated HDL might serve as an additional explanation for the reduced HDL levels in many diabetics.

We have recently isolated autologous HDL from normoglycemic subjects, glucosylated one half together with $NaCNBH_3$ [21], and injected glucosylated and control HDL back into the donors [*Kesäniemi and Ilmonen,* unpubl. data]. In each pair of the studies the glucosylated HDL was cleared slightly faster than the control preparation. The same finding was true in a few patients with noninsulin-dependent diabetes mellitus and in 1 subject with insulin-dependent diabetes. Overall, among all the subjects studied the FCR for the glucosylated HDL was 13% higher than that for the control HDL ($p < 0.05$). Also, the electrophoretic mobility on agarose was faster for the glucosylated than for the control preparation. Thus, the results in the human studies with in vitro glucosylated HDL agree with the data obtained in the animal work [34] even though the differences between the glucosylated and control HDL were not as marked among the human subjects as in the guinea pigs. It is also noteworthy that none of these subjects showed a rapid disappearance of glucosylated HDL from the plasma, a phenomen observed for glucosylated LDL in several diabetics [31].

The findings observed with the in vitro glucosylated HDL suggested us to compare the clearance rate of HDL isolated from a diabetic patient with that isolated from a euglycemic subject. We prepared HDL from a normoglycemic patient and from a patient with insulin-dependent diabetes, labeled them with two different isotopes and injected both preparations into both subjects [*Kesäniemi and Ilmonen,* unpub. data]. Thus. the patients received an injection of autologous and homologous HDL. Interestingly, the HDL isolated from the diabetic patient had a faster mobility on agarose electrophoresis than the HDL from the euglycemic individual. Also the HDL isolated from the diabetic patient (homologous) tended to be cleared slightly faster (3%) than the autologous HDL in the euglycemic patient, but in the diabetic individual no differences between the two HDLs could be observed. However, both HDL preparations were cleared 30% faster in the diabetic patient than in the control subject. When the same set of experiments was repeated in a patient with noninsulin-dependent diabetes and in another normoglycemic subject the glucosylated and control HDL were cleared at an equal rate in both patients. Thus, the differences in the metabolism of HDL in diabetics and normoglycemic subjects may not be closely related to nonenzymatic glucosylation of HDL apoproteins. Several other factors have been reported important for the altered metabolism of HDL in diabetes [5]. At any rate, the enhanced electrophoretic mobility of diabetic HDL on agarose may still indicate changes that can have importance in some functional aspects.

Acknowledgments

Part of these studies were supported by grants from the Sigrid Suselius Foundation and the Foundation for Diabetic Research, Finland.

References

1 Kannel, W.B.; Gordon, T.; Castelli, W.P.: Obesity, lipids and glucose intolerance. The Framingham Study. Am. J. clin. Nutr. 32: 1238–45 (1979).
2 Nikkilä, E.A.; Kekki, M.: Plasma triglyceride transport kinetics in diabetes mellitus. Metabolism 22: 1–22 (1973).
3 Abrams, J.J.; Ginsberg, H.; Grundy, S.M.: Metabolism of cholesterol and plasma triglycerides in nonketotic diabetes mellitus. Diabetes 31: 903–910 (1982).
4 Kissebah, A.H.; Alfarsi, S.; Evans, D.J.; Adams, P.W.: Integrated regulation of very low density lipoprotein triglyceride and apolipoprotein-B kinetics in non-insulin-dependent diabetes mellitus. Diabetes 31: 217–255 (1982).
5 Nikkilä, E.A.: HDL in relation to the metabolism of triglyceride-rich lipoproteins; in Miller, Clinical and metabolic aspects of high-density lipoproteins, pp. 217–245 (Elsevier, Amsterdam 1984).
6 Bradley, R.F.: Cardiovascular disease; in Markle, Diabetes mellitus, pp. 417–477 (Lea & Febiger, Philadelphia 1971).
7 Brownlee, M.; Vlassara, H.; Cerami, A.: Nonenzymatic glycosylation and the pathogenesis of diabetic complications. Ann. intern. Med. 10: 527–537 (1984).
8 Means, G.E.; Chang, M.K.: Nonenzymatic glycosylation of proteins. Structure and function changes. Diabetes 31: suppl. 3, pp. 1–4 (1982).

9   Higgins, P.J.; Bunn, H.F.: Kinetic analysis of the nonenzymatic glycosylation of hemoglobin. J. biol. Chem. *256:* 5205–5208 (1981).
10   Curtiss, L.K.; Witztum, J.L.: A novel method for generating region-specific monoclonal antibodies to modified proteins. Application to the identification of human glucosylated low density lipoproteins. J. clin. Invest. *72:* 1427–1438 (1983).
11   Witztum, J.L.; Mahoney, E.M.; Branks, M.J.; Fisher, M.; Elam, R.; Steinberg, D.: Nonenzymatic glycosylation of low-density lipoprotein alters its biologic activity. Diabetes *31:* 283–291 (1982).
12   Sasaki, J.; Cottam, G.L.: Glycosylation of human LDL and its metabolism in human skin fibroblasts. Biochem. biophys. Res. Commun. *104:* 977–983 (1982).
13   Kim, H.-J.; Kurup, I.V.: Nonenzymatic glycosylation of human plasma low density lipoprotein. Evidence for in vitro and in vivo glucosylation. Metabolism *31:* 348–353 (1982).
14   Gonen, B.; Baenziger, J.; Schonfeld, G.; Jacobson, D.; Farrar, P.: Nonenzymatic glycosylation of low density lipoproteins in vitro. Effects on cell-interactive properties. Diabetes *30:* 875–878 (1981).
15   Sasaki, J.; Cottam, G.L.: Glycosylation of LDL decreases its ability to interact with high-affinity receptors of human fibroblasts in vitro and decreased its clearance from rabbit plasma in vivo. Biochim. biophys. Acta *713:* 199–207 (1982).
16   Goldstein, J.L.; Brown, M.S.: Atherosclerosis: the low density lipoprotein receptor hypothesis. Metabolism *26:* 1257–1275 (1977).
17   Attie, A.; Pittman, R.C.; Steinberg, D.: Hepatic catabolism of low density lipoprotein: mechanisms and metabolic consequences. Hepatology, Baltimore *2:* 269–281 (1982).
18   Mahley, R.W.; Weisgraber, K.H.; Melchior, G.W.; Innerarity, T.L.; Holcombe, K.S.: Inhibition of receptor-mediated clearance of lysine and arginine-modified lipoproteins from the plasma of rats and monkeys. Proc. natn. Acad. Sci. USA *77:* 225–229 (1980).
19   Shepherd, J.; Bicker, S.; Lorimer, A.R.; Packard, C.J.: Receptor-mediated low density lipoprotein catabolism in man. J. Lipid Res. *20:* 999–1006 (1979).
20   Steinbrecher, U.P.; Witztum, J.L.; Kesäniemi, Y.A.: Comparison of glucosylated low density lipoprotein with methylated or cyclohexanedione-treated low density lipoprotein in the measurement of receptor-independent low density lipopotein catabolism. J. clin. Invest. *71:* 960–964 (1983).
21   Kesäniemi, Y.A.; Witztum, J.L.; Steinbrecher, U.P.: Receptor-mediated catabolism of low density lipoprotein in man. Quantitation using glucosylated low density lipoprotein. J. clin. Invest. *71:* 950–959 (1983).
22   Kesäniemi, Y.A.; Vuoristo, M.; Miettinen, T.A.: Metabolism of very low density and low density lipoproteins in patients with coeliac disease; in Carlson, Treatment of hyperlipoproteinemia, pp. 49–53 (Raven Press, New York 1984).
23   Bilheimer, D.B.; Grundy, S.M.; Brown, M.S.; Goldstein, J.L.: Mevinolin and colestipol stimulate receptor-mediated clearance of low density lipoprotein from plasma in familial hypercholesterolemia heterozygotes. Proc. natn. Acad. Sci. USA *80:* 4124–4128 (1983).
24   Slater, H.R.; McKinney, L.; Packard, C.J.; Shepherd,, J.: Contribution of the receptor pathway to low density lipoprotein catabolism in humans. New methods for quantitation. Arteriosclerosis *4:* 604–613 (1984).

25 Sleicher, E.; Deufel, T.; Wieland, O.H.: Nonenzymatic glucosylation of human serum lipoproteins: elevated ε-lysine glycosylated low density lipoprotein in diabetic patients. FEBS Lett. *129:* 1–4 (1981).
26 Steinbrecher, U.P.; Witztum, J.L.: Glucosylation of low density lipoproteins to an extent comparable to that seen in diabetes slows their catabolism. Diabetes *33:* 130–134 (1984).
27 Kissebah, A.H.; Alfarsi, S.; Evans, D.J.; Adams, P.W.: Plasma low density lipoprotein transport kinetics in noninsulin-dependent diabetes mellitus. J. clin. Invest *71:* 655–667 (1983).
28 Chait, A.; Bierman, E.L.; Albers, J.J.: Low density lipoprotein receptor activity in cultured human skin fibroblasts. Mechanism of insulin induced stimulation. J. clin. Invest. *64:* 1309–1319 (1979).
29 Mazzone, T.; Foster, D.; Chait, A.: In vivo stimulation of low density lipoprotein degradation by insulin. Diabetes *33:* 333–338 (1984).
30 Lopes-Virella, M.F.; Wohltmann, H.J.; Mayfield, R.K.; Loadholt, C.B.; Colwell, J.A.: Effect of metabolic control on lipid, lipoprotein, and apolipoprotein levels in 55 insulin-dependent diabetic patients. A longitudinal study. Diabetes *32:* 20–25 (1983).
31 Witztum, J.L.; Steinbrecher, U.P.; Kesäniemi, Y.A.; Fisher, M.: Autoantibodies to glucosylated proteins in the plasma of patients with diabetes mellitus. Proc. natn. Acad. Sci. USA *81:* 3204–3208 (1984).
32 Witztum, J.L.; Steinbrecher, U.P.; Fisher, M.; Kesäniemi, A.: Nonenzymatic glucosylation of homologous low density lipoprotein and albumin renders them immunogenic in the guinea pig. Proc. natn. Acad. Sci. USA *80:* 2757–2761 (1983).
33 Bassiouny, A.R.; Rosenberg, H.; McDonald, T.L.: Glucosylated collagen is antigenic. Diabetes *32:* 1182–1184 (1983).
34 Witztum, J.L.; Fisher, M.; Pietro, T.; Steinbrecher, U.P.; Elam, R.L.: Nonenzymatic glucosylation of high-density lipoprotein accelerates its catabolism in guinea pigs. Diabetes *31:* 1029–1032 (1982).

Y. Antero Kesäniemi, MD, Second Department of Medicine,
University of Helsinki, SF-00290 Helsinki (Finland)

The Lipid Research Clinics Coronary Primary Prevention Trial: Results and Implications

Basil M. Rifkind

National Heart, Lung and Blood Institute, Bethesda, Md., USA

Observational epidemiological studies have established that the higher the plasma total cholesterol or low density lipoprotein cholesterol (LDL-C) the greater the risk of developing coronary heart disease (CHD). The view that LDL-C is intimately involved in atherogenesis and CHD and plays a causal role is strengthened by the findings of many other types of epidemiological studies as well as by animal experimentation, observations on the nature of the atherosclerotic lesion, studies of familial hypercholesterolemias and observations on lipoprotein transport.

It is possible, in most cases, to reduce plasma cholesterol or LDL-C levels substantially by dietary means and, where necessary, by appropriate drug treatment. Many have regarded the evidence linking cholesterol levels to CHD as sufficient to recommend such cholesterol-lowering. However, before such treatment can be widely recommended it is necessary to have direct clinical trial evidence that it is of benefit. Numerous clinical trials of prevention of CHD through cholesterol-lowering using either diet or drug have been conducted. Although many of them have resulted in encouraging findings, none has been judged to be conclusive.

The ideal trial of cholesterol-lowering would be a dietary study. It is generally held that diets high in saturated fat and cholesterol lead to high blood total and LDL-C levels which in turn lead to CHD. However, a definitive study of the effects of altering dietary fat intake is not feasible in view of the inability to double-blind such a study in a large free-living population and the extremely high costs that would be involved. The Lipid Research Clinics Coronary Primary Prevention

Trial (LRC-CPPT) was initiated in 1973 as an alternative test of the efficacy of reducing cholesterol levels.

Participants and Methods

The LRC-CPPT was a multicenter, randomized, double-blind, placebo-controlled clinical trial that tested the efficacy of lowering cholesterol levels for the primary prevention of CHD. The study recruited 3,806 men between the ages of 35 and 59 whose initial plasma cholesterol levels were 265 mg/dl or greater (the 95th percentile for men aged 40–49 years). This was accompanied by an LDL-C level of 190 mg/dl or greater. Men with triglyceride levels averaging greater than 300 mg/dl were excluded. The participants were free of clinical CHD at entry. They were identified through the screening of approximately 480,000 age-eligible men.

To ensure comparability of all data across the 12 participating Lipid Research Clinics over a 10-year period, a common protocol depicting all procedures was strictly adhered to by trained and certified clinic personnel. All aspects of the conduct of the study were carefully monitored. A safety and data-monitoring board periodically reviewed the progress of the study, and especially the possibility of serious side effects.

Prerandomization

During the course of four prerandomization visits many baseline measurements were made. At the second screening visit a moderate plasma cholesterol-lowering diet providing 400 mg cholesterol/day and a P:S ratio of 0.8 was prescribed. It was designed to lower plasma cholesterol levels by 3–5%. Those participants whose LDL-C fell below 175 mg/dl (90th percentile) at the 3rd or 4th screening visit in response to the diet were excluded from the study.

Randomization

At the 5th clinic visit eligible participants were randomly divided into two treatment groups at each of the 12 clinics. The randomization process was successful in producing two essentially identical groups at entry into the study in that, of 83 variables compared at baseline, only 5 showed statistical differences, all of which were small.

Medication

Participants were prescribed either the bile acids sequestrant cholestyramine resin at 24 g/day (6 packets/day) or an equivalent amount of placebo. Those who were unable to take 6 packets/day were prescribed a reduced dose. Compliance to medication was monitored by means of a packet count.

Postrandomized Visits

CPPT participants attended clinic every 2 months for the 7–10 years of follow-up. At the end of the study, contact was made with all of the men who were still living, including any who had discontinued visits during the course of the trial. Thus the vital status was known for all men originally entered into the study. Additionally every man, or a close relative, was questioned before the end of the study regarding previous hospitalizations for CHD or other reasons.

Endpoints

The primary endpoint of the LRC-CPPT, as specified before the study was commenced and around which the sample size estimates were made, was the combination of definite CHD death and/or definite nonfatal myocardial infarction. Strict criteria for these and other endpoints were developed and utilized. Many other coronary and noncoronary endpoints were monitored including all-cause mortality, the development of an ischemic ECG response to a standardized exercise test (positive exercise test) and angina pectoris as determined by the Rose questionnaire.

Adherence

During the first year of the study the mean daily packet count for participants attending clinic was 4.2 in the cholestyramine group and 4.9 in the placebo group. This fell to 3.8 and 4.6, respectively, by the 7th year. Adherence to the diet, as determined by a 24-hour dietary recall, conducted at 6-month intervals, showed no important differences between the two treatment groups.

Cholesterol-Lowering

The initial dietary intervention was associated in both groups with a 3.4% fall in total and a 3.8% fall in LDL-C (fig. 1). The introduction of cholestyramine resin was accompanied by additional falls of 14% in

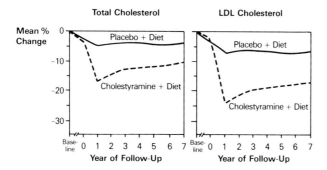

Fig. 1. Mean yearly plasma total cholesterol and LDL-C levels for cholestyramine- and placebo-treated men.

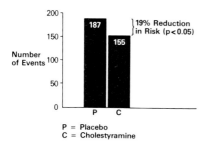

Fig. 2. Comparison of primary endpoints occurring in the cholestyramine- and placebo-traeted groups.

total cholesterol and 21% in LDL-C during the first year of follow-up. Mean total LDL-C levels rose slowly in the cholestyramine-treated men in the succeeding years, probably attributable mainly to the declines in adherence to medication. The cholestyramine group experienced average plasma total cholesterol and LDL-C reductions of 13.4 and 20.3%, respectively, during treatment, which were 8.5 and 12.6% greater than those obtained in the placebo group ($p < 0.001$).

Primary Endpoint

Over the course of the study the cholestyramine group experienced 155 definite CHD deaths and/or definite nonfatal myocardial infarctions, whereas the placebo group had 187 such events (fig. 2). Using the stratified log rank test and taking into account the baseline characteris-

Table I. Definite or suspect primary endpoints and all-cause mortality

Endpoint	Placebo (n = 1,900)		Cholestyramine resin (n = 1,906)		% reduction in risk	z score
	n	%	n	%		
Definite CHD death and/or definite nonfatal myocardial infarction	187	9.8	155	8.1	19	1.92
Definite CHD death	38	2.0	30	1.6	24	–
Definite nonfatal myocardial infarction	158	8.3	130	6.8	19	–
Definite or suspect CHD death or nonfatal myocardial infarction	256	13.5	222	11.6	15	1.80
Definite or suspect CHD death	44	2.3	32	1.7	30	–
Definite or suspect nonfatal myocardial infarction	225	11.8	195	10.2	15	–
All-cause mortality	71	3.7	68	3.6	7	0.42

The 0.05 level, one-sided critical value of the z score adjusted for multiple looks at the data is 1.87.

tics of the participants at entry and the differing lengths of follow-up, we estimated the incidence rate of CHD to be 19% lower in the cholestyramine than in the placebo group (table I). The Z score for this difference was 1.92 ($p < 0.05$) after adjustment for multiple looks at the data and using a one-sided critical value for the Z score. Both fatal and nonfatal categories of the primary endpoint showed corresponding reductions of 24 and 19%, respectively.

Statistical analyses indicated that the benefit of cholestyramine resin treatment could not be attributed to effects in only a small number of clinics. Furthermore, other analyses indicated that it was highly unlikely that the treatment benefit could have arisen from any inequality of the two treatment groups with respect to CHD risk at baseline or from a particular subgroup of LRC-CPPT participants.

Other Cardiovascular Endpoints

Each of the CHD categories with a sufficient number of events to allow a meaningful analysis showed a reduction in incidence similar to that experienced for the primary endpoint. Thus the cholestyramine group showed reductions of 20% ($p < 0.01$) in the incidence of the development of angina, 25% ($p < 0.001$) in the development of an ischemic response to exercise, and 21% ($p = 0.06$) in the incidence of coronary bypass surgery.

All-Cause Mortality

All-cause mortality in the LRC-CPPT was reduced by only 7%. This reflected an increase in deaths not caused by CHD, no differences being observed in deaths from malignant neoplasms or from other medical causes. The only noteworthy difference was 11 deaths from accidents or violence in the cholestyramine group, as compared with 4 in the placebo group. Detailed evaluation of the accidental and violent deaths made it virtually certain or highly unlikely that CHD was the underlying cause of death in any of these cases. Cholestyramine use was also not associated with an increase number of hospitalizations for depression or with the use of antidepressant drugs. The likeliest explanation for the small increase in accidental or violent deaths was believed to be chance.

Assessment for Possible Confounding

A large number of variables, including the several major risk factors for CHD, were monitored during the course of the study. The change from baseline with respect to these was similar in the two treatment groups and did not explain the treatment benefit.

Side Effects and Toxicity

There were no noteworthy differences in nongastrointestinal side effects between the two treatment groups. Gastrointestinal side effects were frequently noted in the placebo- and cholestyramine-treated participants, especially the latter. Constipation and heartburn were especially more frequent in the cholestyramine group, which also reported more abdominal pain, belching, gas and nausea. These were usually not severe and could be dealt with by standard clinical means. They were especially prevalent in the first year but diminished in frequency so that, by the 7th year, approximately equal numbers of cholestyramine- and placebo-treated participants experienced these side effects. Monitoring of a variety of clinical chemical measurements showed some slight changes but none were regarded as of clinical significance.

A large number of categories of hospitalization were monitored during the course of the study. Of these, the only differences with nominal statistical significance were more deviated nasal septa in the cholestyramine group and more operations or procedures involving the nervous system. Although small differences were found in the frequency of diagnoses and procedures involving the gallbladder, none was significant. No deaths attributable to gallbladder disease were recorded.

The number of nonfatal or fatal cases of malignant neoplasms was similar in the two groups. The cholestyramine group had a few more malignant neoplasms in some categories, for example, the buccal cavity and the pharynx, but fewer lung and prostate cancers and melanomas. When the various categories of gastrointestinal cancers were considered together, there were 11 incident cases and 1 fatal case in the placebo group and 21 incident and 8 fatal cases in the cholestyramine group. The number of incident colon cancers in each group was identical.

In general, the numbers of individuals in the various categories are small and too much significance should not be attached. However, in view of the fact that cholestyramine acts in the gastrointestinal tract and that experiments have been reported in which the drug has been found to be a promotor of colon cancer in rats, further follow-up of the LRC-CPPT participants is planned.

Cholesterol-Lowering and Reduction of Risk

A series of analyses was conducted within the cholestyramine-treated group, taking advantage of the variable cholesterol-lowering obtained within that group. Using the Cox proportional hazards analysis it was shown that the greater the adherence to cholestyramine, as measured by the mean daily packet count, the lower the incidence of CHD. Similarly, the greater the reduction in total or LDL-C, the greater the reduction in CHD incidence ($p < 0.001$). When terms for the packet count and the percentage reduction in either total cholesterol or LDL-C were included simultaneously in the model, the term for either expression of cholesterol change remained significant ($p < 0.001$) while the term for packet count was noncontributory ($p > 0.85$). This suggests that the reduction in CHD obtained through cholestyramine therapy was mediated through the drug's effect on cholesterol-lower-

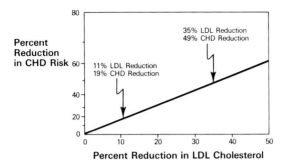

Fig. 3. Relation of reduction in LDL-C to reduction in CHD, as determined by the Cox proportional hazards model.

ing. Furthermore, the analysis indicated that in men sustaining a 25% fall in total cholesterol or 35% in LDL-C levels, typical responses to the full dose (24 g/day) of the resin, CHD incidence was almost half that of men who remained at pretreatment levels (fig. 3).

Consistency with Observational Studies

The placebo-treated participants are, to some extent, comparable to epidemiological observational studies since intervention in this group was minimal. The LDL-C level of placebo participants at entry into the LRC-CPPT was correlated with the subsequent incidence of CHD events; a 22.3 mg/dl decrement in LDL-C, the mean treatment differential actually attained in the study, predicted a 16% reduction in CHD incidence. The predictive power of the placebo group was very close to that of several other observational studies of men in the same range, especially the Framingham Study.

External Consistency with Other Clinical Trials

The relationship between cholesterol-lowering and reduction in risk as seen in the LRC-CPPT was compared with that reported by 10 other trials of cholesterol-lowering in which comparison was possible (fig. 4). Although most of these trials did not report a statistically 'significant' treatment benefit with respect to their predefined primary endpoint, most of them showed beneficial trends. The results of 7 of the 10 studies closely fitted the regression line based on the proportional

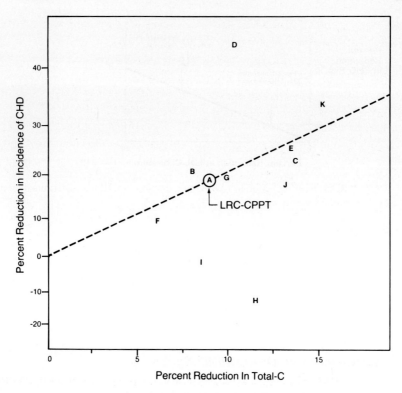

Fig. 4. Comparison of results of 11 cholesterol-lowering trials with experience of the LRC-CPPT cholestyramine-treated group. A = LRC-CPPT; B = WHO Clofibrate Trial; C = Los Angeles Veterans Administration Trial; D = Newcastle Trial; E = Edinburgh Trial; F = Coronary Drug Project (CDP)-Clofibrate Group; G = CDP-Nicotinic Acid Group; H = CDP-Dextrothyroxine Group; I = London Medical Research Council (MRC)-Low Fat Diet; J = London MRC-Soya Bean Oil; K = Oslo Diet-Heart Study.

hazards analysis of CHD incidence within the LRC-CPPT cholestyramine group. The findings also closely agree with the results of a similar analysis previously performed by Peto.

Conclusions

The CPPT findings show that reducing total cholesterol by lowering LDL-C levels diminishes the incidence of CHD morbidity and mortality in men at high risk for CHD because of raised LDL-C levels. These conclusions are based on (1) the significant reduction in the primary endpoint; (2) the corresponding reductions in the secondary endpoints of angina, the development of a positive exercise test, and progression to bypass surgery; (3) the analysis in the cholestyramine group indicating that the reduction in risk was related to the degree of cholesterol-lowering; (4) by the consistent plasma cholesterol-CHD risk relationships as seen when the cholestyramine and placebo groups were compared in the actual clinical trial experiment, within the cholestyramine group or as predicted from the placebo group baseline values, and (5) the consistency of the findings of the LRC-CPPT with the results of many other cholesterol-lowering trials and with what was predicted from observational studies.

Implications of the LRC-CPPT

The most conservative view of the LRC-CPPT would be to restrict its findings to the type of people studied, namely, middle-aged men with cholesterol levels above 265 mg/dl, free of CHD at entry. This, in itself, would have major clinical and public health implications since it involves over 5% (1-2 million) of middle-aged US males. A recent Consensus Development Conference on Lowering Blood Cholesterol to Prevent Heart Disease considered the results of the LRC-CPPT and several other recent studies, together with a variety of other evidence relating cholesterol to CHD. The experts on the Consensus Panel concluded that the data derived from the wealth of congruent results of genetic, experimental pathologic, epidemiological and intervention studies established beyond any reasonable doubt the close relationship between elevated blood cholesterol levels and CHD. They believed that it had been established that lowering definitely elevated blood cholesterol levels (specifically blood levels of LDL-C) would reduce the risk of heart attacks due to CHD. While this has been demonstrated most conclusively in men with elevated blood cholesterol levels, they reported that much evidence justified the conclusion that similar protection would be afforded in women.

The Consensus Development Conference Statement recommended that individuals with 'high-risk blood cholesterol levels' (values above the 90th percentile) be intensively treated by diet and, when necessary, by drugs (table II). It further stated that individuals with 'moderate-risk blood cholesterol levels' (values between the 75th and 90th percentiles)

Table II. Cholesterol values for selecting adults at moderate and high risk, requiring treatment

Age group	Moderate risk		High risk	
	mg/dl	mM	mg/dl	mM
20–29	>200	5.17	>220	5.69
30–39	>220	5.69	>240	6.21
40+	>240	6.21	>260	6.72

Source: Consensus statement on lowering blood cholesterol to prevent heart disease, vol. 5, No. 7, Dec. 1984 [reprinted in the *J. Am. med. Ass.*, April 12, 1985].

be treated intensively by dietary means. It was also recommended that all Americans cut their total and saturated fat and dietary cholesterol intake in order to lower their average plasma cholesterol levels.

Adoption of these measures has the potential to produce a very marked reduction in CHD risk. Taken in conjunction with control of other major coronary risk factors such as cigarette smoking and hypertension, these measures make it likely that the recent marked decline in CHD mortality, which the United States has experienced, will be maintained together with reductions in many of the morbid consequences of atherosclerotic disease.

Basil M. Rifkind, MD, National Heart, Lung and Blood Institute, Federal 401, Bethesda, MD 20205 (USA)

Chlorophenoxyisobutyric Acid Derivatives and Apolipoprotein B Metabolism

James Shepherd, Christopher J. Packard[1]

Department of Biochemistry, Royal Infirmary, Glasgow, UK

Clofibrate, the ethyl ester of *p*-chlorophenoxyisobutyric acid (CPIB), is one of the longest serving hypolipidaemic agents in clinical use. It was first shown to lower plasma cholesterol concentrations in rats [1], and also has a similar action in man [2]. Its combined clinical utility and patient acceptability has led a number of pharmaceutical companies to formulate derivatives (fig. 1) which seem to function in the same way, although with a few, possibly important qualitative differences. CPIB derivatives exhibit a wide range of metabolic activities but are generally prescribed because of their effects on plasma triglyceride and cholesterol. Characteristically, triglyceride levels fall due to a reduction in circulating very low density lipoprotein (VLDL) which is evident within 2–5 days of initiating therapy. Concomitantly, there is generally a decrease in cholesterol levels in whole plasma and in the low density lipoprotein (LDL) fraction although this is less predictable and dramatic. Prolonged experience with the drugs has revealed the reason for such variability. When they are administered to severely hypertriglyceridaemic subjects there may be a rise in the initially subnormal plasma LDL levels of these patients: conversely, treatment of normotriglyceridaemic hypercholesterolaemic subjects produces a fall [3]. This paradoxical response has provided important clues to the mechanisms of action of the drugs (see below).

---

[1] We acknowledge the secretarial expertise of *Joyce Pollock*. This work was performed during the tenure of grants from the Scottish Home and Health Department (K/MRS/50/C429) and the Scottish Hospital Endowments Research Trust (HERT 673).

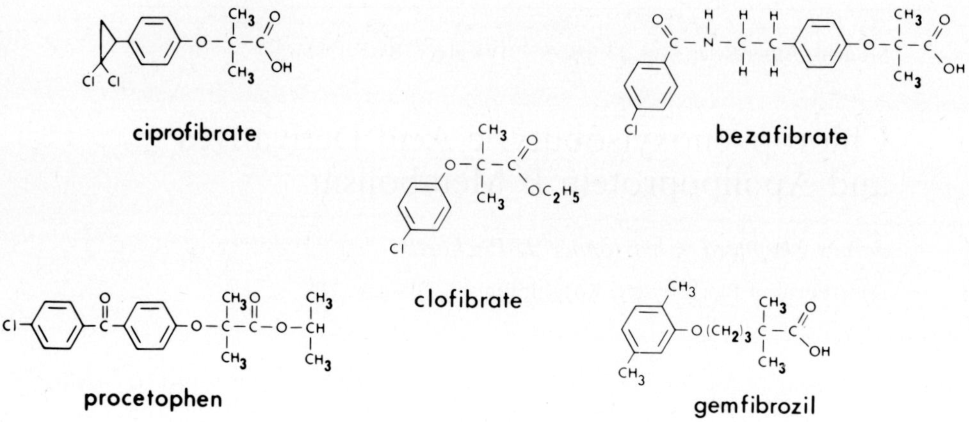

Fig. 1. Clofibrate and some of its clinically useful derivatives.

Actions of Clofibrate Analogues on Plasma Lipids and Lipoproteins

Although the fibrates are known to exert a wide variety of pharmacologic effects, most attention has been focused on their influence on plasma lipid and lipoprotein metabolism. In particular, intensive efforts have been made to unravel the changes which they induce in the turnover of apolipoprotein B (Apo B), the main structural polypeptide of both VLDL and LDL. Despite this, it is only recently that we have begun to grasp their involvement in the modulation of lipoprotein kinetics. Figure 2 outlines current concepts of their actions on VLDL, intermediate density lipoprotein (IDL) and LDL. They appear to exert multiple, possibly independent influences on triglyceride- and cholesterol-rich Apo B-containing particles.

Fibrates and Triglyceride-Rich Particles

As has already been noted, clofibrate analogues produce a substantial reduction in plasma triglyceride and VLDL levels. This may derive from limitation of their synthesis or from an increase in their catabolism, and there is evidence that both mechanisms may be involved. Several lines of evidence support the concept that substrate availability for triglyceride synthesis is diminished during therapy. First (fig. 3), clofibrate has been shown to suppress the increase in plasma-free fatty acids which follows the administration of adrenaline to both animals and man [4]. Secondly, it limits the availability of cyclic AMP [5], the

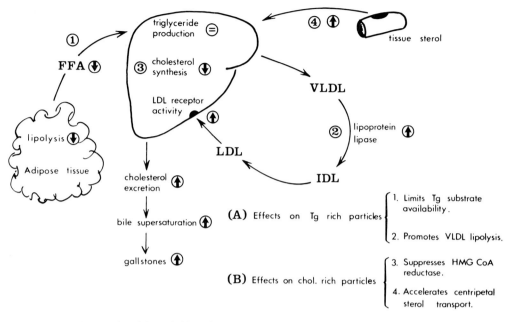

Fig. 2. Postulated hypolipidaemic actions of the CPIB analogues.

intracellular second messenger involved in the lipolytic process. These two events presumably contribute to the reduced flux of plasma-free fatty acids which is associated with clofibrate treatment [6]. By inference, such a situation might conspire to limit hepatic triglyceride synthesis by restricting delivery of exogenous fatty acid substrate to that organ. However, *Barter* et al. [7] have shown that only about 50% of the free fatty acids required for triglyceride production come from the plasma, the remainder presumably being derived from a hepatic pool. Here again, however, the fibrates may have a limiting action. In the liver they are known to suppress the activity of acetyl-CoA carboxylase [8] and to activate α-glycerophosphate dehydrogenase [9], effectively inhibiting endogenous fatty acid synthesis and limiting glycerol availability by promoting its oxidation. These events, then, would conspire to suppress triglyceride synthesis, a credible explanation for the reduction in plasma triglyceride levels which follow fibrate administration.

Despite the plausibility of the above arguments, many investigators now feel that clofibrate and its analogues exert their primary action on

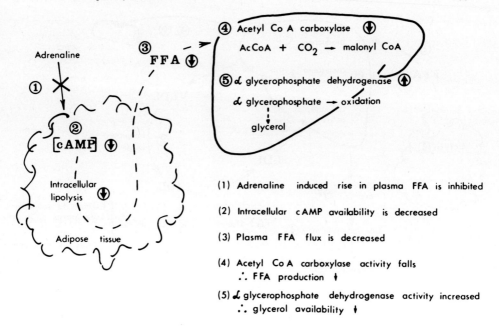

Fig. 3. Effects of CPIB derivatives on triglyceride synthesis.

triglyceride-rich particles by accelerating their rate of utilization in peripheral tissues. This view was first advanced by *Wolfe* et al. [10] and gained support from the observations that treatment with the drug increased postheparin lipoprotein lipase activity [11, 12] and stimulated the fractional removal rate of Intralipid from the plasma [11, 12]. The response to fibrate therapy of hepatic triglyceride lipase, the other major intravascular lipolytic enzyme, is variable.

Fibrates and Cholesterol-Rich Particles

Although the hypocholesterolaemic action of clofibrate is usually much less dramatic than its influence on plasma triglyceride, in practical clinical terms it may have much greater significance since plasma cholesterol is strongly correlated with coronary heart disease. The balance between plasma LDL cholesterol and intracellular sterol pools is maintained by the activity of the high affinity LDL receptor pathway which is stimulated to draw upon circulating lipoproteins in times of cellular cholesterol need. Such a situation might arise if de novo choles-

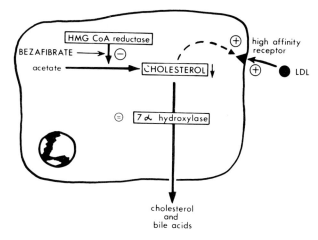

Fig. 4. Postulated hypocholesterolaemic action of bezafibrate. By suppressing endogenous cholesterol synthesis through its inhibitory action on HMG-CoA reductase, the drug may promote LDL assimilation via high affinity receptors on cell membranes.

terologenesis is inhibited, particularly in the liver. The fibrates are known to inhibit [13] 3-hydroxy-3-methylglutaryl-coenzyme A reductase (HMG-CoA reductase), the rate-limiting enzyme for sterol synthesis. This mechanism (fig. 4) has been used to explain the observation that bezafibrate therapy in man accelerates receptor-mediated catabolism of LDL [14], although it is still not clear whether HMG-CoA reductase activity in man is suppressed by the drugs.

In addition to their actions on LDL, clofibrate derivatives variably influence the level of HDL in the plasma. In many hypertriglyceridaemic subjects, whose HDL levels are characteristically low, treatment raises the cholesterol content of this fraction. It is possible that this effect, coupled with the increased delivery of LDL to the liver may provide a means of accelerating centripetal sterol transport to the liver for ultimate excretion.

CPIB Derivatives and Their Effects on Apolipoprotein B Metabolism

In a recent series of studies we have examined the metabolism of apolipoprotein B in subfractions of VLDL and LDL before and during

Table I. Effects of bezafibrate on ApoB kinetics in subjects with hypertriglyceridaemia

Subjects (n = 6)	VLDL		IDL		LDL	
	pool size mg/dl	fractional clearance rate pools/day	pool size mg/dl	fractional clearance rate pools/day	pool size mg/dl	fractional clearance rate pools/day
Control	10.5±10.8	7.0± 7.5	30.0± 6.2	1.23±0.55	55±28	0.47±0.25
Bezafibrate	3.1± 3.6	22.9±24.0	38.9±11.5	0.98±0.38	76±22	0.35±0.12
Pair difference t test	<0.05	<0.05	<0.05	NS	NS	<0.05

fibrate administration. Hypertriglyceridaemia is characterized by the accumulation of large triglyceride-rich VLDL particles (Sf 100–400) in the circulation. These were isolated, radiolabeled, and the transit of their Apo B along the metabolic cascade to IDL (Sf 12–100) and LDL (Sf 0–12) examined in a group of hypertriglyceridaemic individuals. In common with our experience in normal subjects we found that these patients produced remnant particles of intermediate (Sf 12–100) density which thereafter were removed slowly from the circulation without appearing in LDL. When the patients were given bezafibrate the plasma concentration of the large triglyceride-rich particles decreased by enhancement of their catabolism (table I). Synthesis of VLDL Apo B into this density interval was unaltered. This acceleration of lipolysis appeared to expand the remnant IDL (Sf 12–100) population whose catabolism remained the same. Plasma LDL concentrations also tended to increase, particularly in the grossly hypertriglyceridaemic individuals with initially low values for this parameter. Parameter analysis of the data suggested that the effect came from a reduction in the fractional rate of catabolism of the LDL. The phenomenon was examined in greater detail in a study which is reported elsewhere in this volume [15].

References

1 Thorp, J.M.: An experimental approach to the problem of disordered lipid metabolism. Atherosclerosis 3: 351–360 (1963).
2 Oliver, M.F.: Reduction of serum lipid and uric acid levels by an orally active andosterone. Lancet i: 1321–1323 (1962).

3 Olsson, A.G.; Rossner, S.; Walldius, G.; Carlson, L.A.; Lang, P.D.: Effects of BM 15.075 on lipoprotein concentrations in different types of hyperlipoproteinaemia. Atherosclerosis 27: 279–287 (1977).
4 Hunninghake, D.B.; Azarnoff, D.L.: Clofibrate effect on catecholamine induced metabolic changes in humans. Metabolism 17: 588–595 (1968).
5 Carlson, L.A.; Walldius, G.; Butcher, R.W.: Effect of chlorophenoxyisobutyric acid (CPIB) on rat-mobilising lipolysis and cyclic AMP levels in rat epididymal fat. Atherosclerosis 16: 349–357 (1972).
6 Rifkind, B.M.: Effect of CPIB ester on plasma free fatty acid levels in man. Metabolism 15: 673–675 (1966).
7 Barter, P.J.; Nestel, P.J.; Kevin, F.C.: Precursor of plasma triglyceride fatty acid in humans. Effect of glucose consumption, clofibrate administration and alcoholic fatty liver. Metabolism 21: 117–124 (1972).
8 Maragandakis, M.E.; Hankin, H.: On the mode of action of lipid lowering agents. V. Kinetics of the inhibition in vitro of rat acetyl CoA carboxylase. J. Biochem. 246: 348–354 (1971).
9 Tarentino, A.L.; Richert, D.A.; Westerfield, W.W.: The concurrent induction of hepatic α-glycerophosphate dehydrogenase by thyroid hormone. Biochim. biophys. Acta 124: 295–309 (1966).
10 Wolfe, B.M.; Kane, J.P.; Havel, R.J.; Brewster, H.P.: Mechanism of the hypolipidemic effect of clofibrate in postabsorptive man. J. clin. Invest. 52: 2146–2159 (1973).
11 Lithell, H.; Boberg, J.; Hellsing, K.; Lundqvist, G.; Vessby, B.: Increase of lipoprotein lipase activity in human skeletal muscle during clofibrate administration. Eur. J. clin. Invest. 8: 67–76 (1978).
12 Vessby, B.; Lithell, H.; Hellsing, K.; Ostlund-Lindqvist, A-M.; Gustafsson, I.B.; Boberg, J.; Ledermann, H.: Effects of bezafibrate on the serum lipoprotein composition, lipoprotein triglyceride removal capacity and the fatty acid composition of the plasma lipid esters. Atherosclerosis 37: 257–269 (1980).
13 Berndt, J.; Gaumert, R.; Still, J.: Mode of action of the lipid lowering agents clofibrate and BM 15.075 on cholesterol biosynthesis in rat liver. Atherosclerosis 30: 147–152 (1978).
14 Stewart, J.M.; Packard, C.J.; Lorimer, A.R.; Boag, D.; Shepherd, J.: Effects of bezafibrate on receptor mediated and receptor independent low density lipoprotein catabolism in type II hyperlipoproteinaemic subjects. Atherosclerosis 44: 355–365 (1982).
15 Packard, C.J.; Caslake, M.J.; Shepherd, J.: Effects of fenofibrate on receptor mediated and receptor independent low density lipoprotein catabolism in hypertriglyceridaemic subjects. Monogr. Atheroscler., vol. 13, pp. 142–144 (Karger, Basel 1985).

James Shepherd, MD, Department of Biochemistry, Royal Infirmary,
Glasgow G4 0SF (UK)

Hyperglycaemia as a Risk Factor for Coronary Heart Disease

Frederick H. Epstein

Institute for Social and Preventive Medicine, University of Zürich, Switzerland

It is over 30 years ago since Sir George Pickering made the fundamentally important statement that hypertension is not a distinct entity but merely the upper end of a continuous distribution, blood pressure levels being determined over the whole range by the interaction of multiple genetic and environmental influences. The importance of this view goes far beyond the concept of hypertension and has particular bearing on the nature of diabetes. Therefore, in the early 1960s, the hypothesis was tested in the Tecumseh Study that the association between diabetes and atherosclerosis might also be quantitative in the sense that increasing degrees of hyperglycaemia, short of clinical diabetes, might increase the risk of coronary disease. This was, indeed, the case in terms of prevalence data [4]. In Tecumseh, we measured blood glucose 1 h after a standard load but later, using incidence data, it was shown in Framingham that a casual blood glucose determination was likewise related to coronary heart disease risk in a curvilinear fashion [15]. In the 1970s, doubt was cast on regarding hyperglycaemia, or rather glucose intolerance, as an independent risk factor and, in order to get more decisive information, the international collaborative group study was organized by *Stamler* [16], based on a retrospective analysis of already existing incidence data from a variety of epidemiological investigations. In this brief review, the data bearing on this question will be summarized.

It is useful to divide the studies into those in which a glucose loading test was used (table I) and the remainder, based on measurements of casual glucose levels, obtained whenever the participant hap-

Table I. Coronary heart disease and glucose intolerance – 13 prospective studies with loading test

Correlation		Study	Reference
univariate	multivariate		
+	+	Whitehall 1983	6
+	+	Hawai (Japanese) 1982	20
+	+	Chicago Gas Co.	16
+	+	Basel (exercise ECG)	16
+	+	Glostrup (exercise ECG)	16
+	–	Bedford 1982[1]	8
+	–	Helsinki Police	16
–		Paris Police	16
–		Helsinki Social Insurance	16
–		Chicago Industry	16
–		Chicago Western Electric	16
–		Busselton	16
–		Tecumseh 1981[1]	12

[1] Men and women (all other studies: only men).

pened to come for examination (table II). All the studies reviewed are prospective, omitting prevalence studies which are less telling. Concerning the 13 studies with a loading test (table I), 9 are full members of the International Collaborative Group [16]. In the first block of 5 studies, there is a positive association between glucose intolerance both on univariate and multivariate analysis. Multivariate analysis is essential because glucose intolerance is correlated with several other risk factors, especially blood pressure [17], and the association could be mediated by these correlated variables. This is not the case for the first 5 studies. In 2 further studies (Bedford, England, and the Helsinki Police Force), glucose intolerance has no independent predictive power, since the correlations disappear on multivariate analysis. In the remaining 6 studies, glucose intolerance and coronary heart disease are not correlated. On the balance, therefore, a correlation is present in 5 of the 13 studies, borderline in 2 and absent in 6 studies. It must be recognized that it is difficult to compare studies which used different

Table II. Coronary heart disease and glucose intolerance – 6 prospective studies with casual blood sugar test

Correlation		Study	Reference
univariate	multivariate		
+	no information	Framingham 1971[1]	15
+	no information	Israel 1973	11
+	+ (not significant)	Puerto Rico 1983	2
+	+ (significant for men only)	S. California 1984[1]	1
–		Renfrew (Scotland)	16
–		Chicago Gas Co.	16

[1] Men and women (all other studies: only men).

methods and criteria though an effort was made in the International Collaborative Study to take account of this problem. In 2 of the studies (Basel and Glostrup), exercise electrocardiograms were included in making a diagnosis of coronary heart disease.

Turning to 6 further studies (table II) in which the diagnosis of glucose intolerance was based on a casual blood sugar test, there was a correlation between the disease and the test result in 4 of the investigations but in only 2 was a multivariate analysis carried out. The latter gave positive correlations but they were significant for men only in 1 of the studies. There was no positive correlation in the 2 other studies. One of these is the Chicago Gas Company Study in which a positive result on multivariate analysis was obtained with a loading test (table I).

It is apparent from these results (table I, II) that glucose intolerance is not an established risk factor like serum lipids, blood pressure and smoking. On the other hand, a sufficient proportion of studies give positive findings to justify the view that there is a need for further investigation. As a matter of fact, the lack of uniformity in the results is, in retrospect, not surprising because glucose intolerance is not directly linked to the mechanisms of atherosclerosis and coronary heart disease like the 3 major risk factors mentioned. Instead, glucose intolerance had always been considered a marker for some other metabolic disturbance, not necessarily strongly associated with hyperglycaemia but more closely linked with the mechanisms of the disease [5]. The most obvious candidate for such a role is insulin. There are now 3 prospective studies in which elevated serum insulin has been shown to be a risk factor for coronary heart disease [3, 13, 18]. Hyperinsulinaemia rather than lack of insulin is a characteristic of a considerable number of latent and manifest diabetics during some stages of disease development and there is considerable evidence that insulin is atherogenic [14]. More generally, there is a need to look more specifically at the intercorrelations between glucose intolerance and possible or established precursors of atherosclerotic disease. As an example, it is not known whether or to what extent low HDL-cholesterol levels which have been shown to be associated with diabetes in some but not all studies, are a feature of impaired glucose tolerance, short of diabetes. Likewise, several measures of haemostatic function, particularly fibrinogen, are deranged in clinical diabetes [7], but there are no corresponding data applying to latent diabetes. There are now 3 prospective studies in which raised serum fibrinogen is an independent predictor of coronary heart disease [9, 10, 19]. It may be hoped that such studies will eventually permit the identification of factors which are more or less closely linked to hyperglycaemia but have a more direct relationship to the mechanisms of atherosclerosis and its clinical consequences.

The answer to the question might not only provide some new insights into the precursors and causes of atherosclerosis but has preventive implications. Manifest diabetes is unquestionably a risk factor for coronary heart disease but, despite its importance as a major chronic disease, it is not common enough to be responsible for more than a relatively small proportion of all myocardial infarctions and sudden deaths in the community. Glucose intolerance, however, is more prevalent and the risk attributable to it could be more appreciable. It was, indeed, the initial hope that reducing the frequency of glucose intolerance in the population might have a considerable impact on reducing the burden of coronary heart disease in the population. While this hope has not been fulfilled in terms of present knowledge, it remains likely that some of the interrelationships between glucose intolerance and potential mechanisms of atherosclerosis may yet provide an additional and new handle on disease prevention.

References

1   Barrett-Connor, E.; Wingard, D.L.; Criqui, M.H.; Suarez, L.: Is borderline hyperglycemia a risk factor for cardiovascular death? J. chron. Dis. *37:* 773–779 (1984).
2   Cruz-Vidal, M.; Garcia Palmieri, M.R.; Costas, R., Jr., et al.: Abnormal blood glucose and coronary heart disease. Diabetes Care *6:* 556–561 (1983).
3   Ducimetière, P.; Eschwege, E.; Papoz, L.; Richard, J.L.; Claude, J.R.; Rosselin, G.: Relationship of plasma insulin levels to the incidence of myocardial infarction and coronary heart disease mortality in a middle-aged population. Diabetologia *19:* 205–210 (1980).
4   Epstein, F.H.; Ostrander, L.D., Jr.; Johnson, B.C., et al.: Epidemiological studies of cardiovascular disease in a total community – Tecumseh, Michigan. Ann. intern. Med. *62:* 1170–1187 (1965).
5   Epstein, F.H.: 'Hyperglycemia' – a risk factor in coronary heart disease. Circulation *36:* 609–619 (1967).
6   Fuller, J.H.; Shipley, M.J.; Rose, G.; Jarrett, R.J.; Keen, H.: Mortality from coronary heart disease and stroke in relation to degree of glycaemia: the Whitehall Study. Br. med. J. *ii:* 867–870 (1983).
7   Jarrett, J.: Diabetes and the heart: coronary heart disease. Clin. Endocrinol. Metab. *6:* 389–402 (1977).
8   Jarrett, R.J.; McCartney, P.; Keen, H.: The Bedford Survey: ten-year mortality rates in newly diagnosed diabetics, borderline diabetics and normoglycaemic controls and risk indices for coronary heart disease in borderline diabetics. Diabetologia *22:* 79–84 (1982).
9   Kannel, W.B.: Personal communication (1985).
10 Meade, T.M.; North, W.R.S.; Chakrabarti, R., et al.: Haemostatic function and cardiovascular death: early results of a prospective study. Lancet *i:* 1050–1054 (1980).
11 Medalie, J.H.; Kahn, H.A.; Neufeld, H.N.; Riss, E.; Goldbourt, U.: Five-year myocardial infarction incidence. II. Associations of single variables to age and birth-place. J. chron. Dis. *26:* 329–349 (1973).
12 Ostrander, L.D., Jr.; Lamphier, D.E.; Carman, W.J.; Williams, G.W.: Blood glucose and risk of coronary heart disease. Arteriosclerosis *1:* 33–37 (1981).
13 Pyörälä, K.; Savolainen, E.; Lehtovirta, E., et al.: Glucose tolerance and coronary heart disease: Helsinki policemen study, J. chron. Dis. *32:* 729–745 (1979).
14 Stout, R.W.: Blood glucose and atherosclerosis. Arteriosclerosis *1:* 227–234 (1981).
15 Kannel, W.B.; Gordon, T.: The Framingham Study: an epidemiological investigation of cardiovascular disease, sect. 27: Coronary heart disease, atherothrombotic brain infarction, intermittent claudications – a multivariate analysis related to their incidence, 16-year follow-up. No. 1740-0320 (US Government Printing Office, Washington 1971).
16 The International Collaborative Group: Joint discussion. J. chron. Dis. *32:* 829–837 (1979).
17 Vaccaro, O.; Rivellese, A.; Riccardi, G., et al.: Impaired glucose tolerance and risk factors for atherosclerosis. Arteriosclerosis *4:* 592–597 (1984).
18 Welborn, T.A.; Wearne, K.: Coronary heart disease incidence and cardiovascular

mortality in Busselton with reference to glucose and insulin concentrations. Diabetes Care 2: 154–160 (1979).
19 Wilhelmsen, L.; Svärdsudd, K.; Korsan-Bengtsen, K., et al.: Fibrinogen as a risk factor for stroke and myocardial infarction. New Engl. J. Med. 311: 501–505 (1984).
20 Yano, K.; Kagan, A.; McGee, D.; Rhoads, G.G.: Glucose intolerance and nine-year mortality in Japanese men in Hawai. Am. J. Med. 72: 71–80 (1982).

Prof. Dr. F.H. Epstein, Lindenstrasse 37, CH-8008 Zürich (Switzerland)

The Diabetes Intervention Study (DIS): A Cooperative Multi-Intervention Trial with Newly Manifested Type II Diabetics: Preliminary Results

M. Hanefeld, J. Schulze, S. Fischer, U. Julius, H. Schmechel, H. Haller, and The DIS Group Dresden, GDR

Medical Clinic, Medical Academy 'Carl Gustav Carus', Dresden, GDR

Atherosclerotic macroangiopathy is the major cause of death in type II diabetics. So far, no prospective intervention study exists in this largest subgroup of diabetics aimed at the primary or secondary prevention of mortality and morbidity due to coronary heart disease (CHD), stroke and peripheral arterial disease. The Diabetes Intervention Study (DIS) is a multicentre, multi-intervention trial with newly manifested unselected type II diabetics classified as 'diet satisfactory'. The main objectives are: (1) basically epidemiologic data on family background, prevalence and incidence of coronary risk factors; (2) factors influencing the course of diabetes, (3) prevalence and incidence of macro- and microangiopathy; (4) influence of risk factor profile on the development of vascular complications; (5) effect of intervention measures upon course of diabetes and vascular complications.

Material and Methods

Based on centralized diabetes registration and care in the GDR, any patient with newly manifested diabetes in the area of the cooperating diabetes clinics was screened (fig. 1). Sixteen diabetic clinics distributed over rural and urban areas of the GDR are involved in the study (DIS Group). The criteria for manifest diabetes used are given in figure 1.

Screening Phase: This period lasted on average 6 weeks and included three visits (0, 4th, 6th weeks). The only treatment during this time was the usual diabetes diet except for patients with type I diabetes or other essential indications for insulin. A patient was grouped as 'diet satisfactory' if his pp blood glucose was < 250 mg/dl and urine sugar < 20 g/24 h at the 2nd and 3rd visit (3 out of 4 values). Admission criteria were

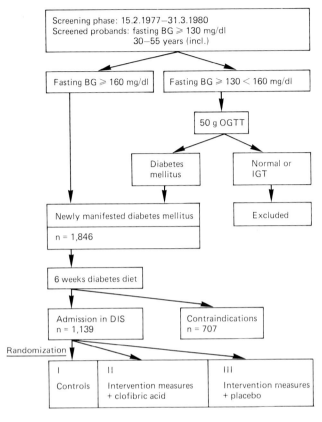

Fig. 1. DIS schedule.

furthermore: pre-existing manifest macroangiopathy (table I), severe life-limiting diseases, clofibrate and anticoagulant intake prior to entry. If no contraindications existed the patients were asked for their consent to participate in the study. The examination program is given in table II. The term 'CHD' refers to ECG changes according to Minnesota codes 1.1–1.3, 4.1–4.3, 5.1–5.3 or 7.1. In brief, the intervention comprises the following measures: (1) 'prudent' lipid-lowering diet; (2) overweight reduction; (3) antismoking education; (4) physical training; (5) standardized management of hypertension; (6) clofibric acid (1.6 g/day) versus placebo.

Table I. Contraindications (%) for DIS participation

Contraindication	%
Myocardial infarction	2.5
Stroke	1.3
Gangrene of the foot	0.3
Other severe diseases	8.8
Indication for oral antidiabetics or insulin	18.8
Not willing or able to participate	13.8

Table II. Examination program

Every 4 months	weight, blood pressure, blood glucose urine sugar (24 h), triglycerides, cholesterol
Annually	interval medical history, short physical examination, diet protocol, physical activity score, smoking (g/day tobacco), resting ECG, routine laboratory tests
0, 3rd, 5th year	medical history, X-ray of the chest, fundus oculi (in subgroups including fluorescence angiography), oscillography
0, 5th year	ergometry, X-ray of the pelvis and foot
5th year	apoprotein B, HDL cholesterol, LDL cholesterol, HbA_1, 75 g OGTT with IRI, in insulin-treated patients glucagon test with C-peptide, gas chromatography: fatty acid pattern of serum triglycerides, clofibric acid blood concentration

Results and Discussion

Indications for antidiabetics, nonacceptance, and pre-existing macroangiopathy were the major causes for exclusions after the screening phase (table I). The prevalence of myocardial infarction was significantly higher as in the general population of this age range in Dresden which is below 1%. As shown in table III, patients typed as 'diet satisfactory' were more overweight and exhibit lower initial blood glucose levels than their counterparts who needed antidiabetics. The diet group lost significantly more weight and reached better glucose levels than those with antidiabetics (table IV). The family history shows a clustering of risk factors and exaggerated prevalence of vascular complications among the parents of our type II diabetics (table V). It

Table III. Baseline data of the screened diabetes population ($\bar{x} \pm SEM$)

	Total	Contra-indications	Diet not satisfactory	Admitted
Height, cm	167.8 ± 8.5	167.5 ± 8.7	167.7 ± 9.0	168.8 ± 8.4
Weight, kg	84.7 ± 15.5	82.3 ± 16.2	79.6 ± 16.1	86.2 ± 14.8
Weight index	125.8 ± 22.6	122.8 ± 23.2	118.8 ± 23.7	127.7 ± 21.9
Age, years	46.4 ± 6.4	46.4 ± 7.3	44.9 ± 8.4	46.4 ± 5.7
Fasting blood glucose, mg/dl	204.2 ± 75.3	229.9 ± 92.3	270.7 ± 99.2	188.3 ± 57.0
Urine sugar, g/24 h	27.8 ± 42.6	38.7 ± 50.2	64.8 ± 56.0	21.5 ± 36.0

Table IV. Changes in body weight, blood glucose and urine sugar excretion during the screening phase

		Total	Contra-indications	Diet not satisfactory	Admitted
PP Blood glucose, mg/dl	2nd visit	155.9 (−47.6)	164 (−64.4)	179.9 (−87.5)	151 (−37.5)
	3rd visit	148.4 (−55.7)	155.6 (−74.9)	165.1 (−103.6)	144.1 (−44.4)
Urine sugar, g/24 h	2nd visit	5.4	9.4	16.4	3
	3rd visit	3.4	6.1	10.4	1.9
Weight, kg	2nd visit	82.9 (−1.8)	80.9 (−1.4)	78.8 (−0.8)	84.2 (−2.0)
	3rd visit	81.9 (−2.8)	80 (−2.3)	78.6 (−1)	83 (−3.2)

Table V. Coronary risk factors (%) and atherosclerotic diseases in the familial medical history of the DIS patients

Disease	Mother (n = 558)	Father (n = 706)	Sibs (n = 156)
Obesity	46.4	24.8	35.5
Diabetes	24.1	11.2	11.1
Hyperlipoproteinaemia	0.7	0.7	1.1
Gout	1.6	1.1	0.7
Myocardial infarction	4.5	9.2	2.3
Hypertension	21.1	9.6	6.9
Stroke	9.8	9.4	1.9
Peripheral arterial disease	6.0	4.2	1.6

Table VI. Causes of death among first-degree relatives of the DIS patients

Disease	Mother (n=558)	Father (n=706)	Sibs (n=156)	p
Diabetes	5.0	1.0	1.3	0.01
Hypertension	0.5	0.1	0	n.s.
CHD	3.4	3.3	1.3	n.s.
Stroke	14.0	7.5	3.8	0.01
Myocardial infarction	5.6	10.9	3.8	0.01
'Atherosclerosis'	4.3	2.7	0	n.s.

Table VII. Clinical findings at entry (%)

Finding	%
X-ray of the thorax	
Left heart hypertrophy	4
Left heart failure	0.4
Peripheral arterial disease	1.7
CHD	
Resting ECG	M 8.0
	F 21.5
Submaximal exercise	M 14.6
Load ECG (n=796)	F 32.0

Table VIII. Prevalence of coronary risk factors (%) among DIS patients and the general population (Dresden study, n=1,216)

	Type II diabetics	Population	Limits
Hyperlipoproteinaemia	17.6	7.6	triglyceride ⩾250 mg/dl and/or cholesterol ⩾300 mg/dl
Hypertriglyceridaemia	11.3	3.4	
Hypercholesterolaemia	3.5	3.7	
Mixed HLP	2.8	0.5	
Hypertension	53.0	17.3	blood pressure ⩾160/95 mm Hg and/or antihypertensive drugs
Smoking	34.0	30.3	tobacco ⩾1 g/day
Obesity	49.0	8.2	IBWI M >1.2 / F >1.3

seems of particular interest that diabetes is twice as frequent among mothers than among fathers. The same applies for hypertension and obesity. Obviously intrauterine environmental factors have a negative effect on proneness for diabetes. The importance of familial factors is emphasized by the causes of death among the parents (table VI). Thus, the family history reveals that 'the shadow of the past' may play an important role for the cardiovascular excess mortality among type II diabetics.

Our data at entry confirm both even presumably mild ('diet satisfactory') newly manifested diabetics exhibit multiple clinical complications (table VII) and they suffer from a clustering of risk factors (table VIII).

Prof. Dr. sc. M. Hanefeld, Medical Clinic, Medical Academy 'Carl Gustav Carus', Fetscherstrasse 74, GDR-8019 Dresden (GDR)

Analysis of Serum Lipoproteins in Insulin-Dependent (Type I) and Noninsulin-Dependent (Type II) Diabetes mellitus[1]

P. Weisweiler, W. Merk, P. Schwandt[2]

Second Medical Department, Grosshadern Clinic, University of Munich, FRG

Introduction

Serum lipoprotein lipid concentrations are altered in diabetes mellitus with associated differences in apolipoprotein levels in diabetics versus control subjects. These differences may account for the increased frequency for atherosclerotic vascular diseases in diabetes mellitus [1–3]. Increased levels of very low-density and low-density lipoproteins (VLDL and LDL) and a decreased concentration of high-density lipoproteins (HDL) have been frequently reported (table I). However, others reported different findings dependent on the fact, whether insulin-dependent diabetics (IDD, type I) or noninsulin-dependent diabetics (NIDD, type II) have been investigated. Apolipoproteins B and E are major apolipoproteins of LDL and of cholesterol-enriched VLDL that are considered atherogenic [21, 22]. Apolipoprotein A-I is the major apolipoprotein of HDL that are held to be an antiatherogenic factor [21, 23]. It has been suggested that apolipoproteins may be better predictive values than lipids for cardiovascular disease [24]. The purpose of our study was, therefore, to provide serum lipoprotein lipid and apolipoprotein characteristics of controls, IDD (type I), and NIDD (type II) subjects and to evaluate possible differences in the serum lipoprotein pattern between both diabetic groups.

---

[1] Supported by the Deutsche Forschungsgemeinschaft (We 955/1–3).
[2] The authors wish to thank Dr. *F. Dati* (Behringwerke, Marburg, FRG) for technical help, Miss *C. Friedl* and Miss *M. Ungar* for their excellent laboratory work.

Materials and Methods

Patients (n = 49) and normal subjects (n = 13) were recruited from the outpatient clinic. Medical histories were reviewed to obtain the present clinical status of the diabetic subjects, 18 of whom were classified as IDD (type I) subjects, 19 as NIDD (type II) subjects without sulphonylureas, and 12 as NIDD (type II) subjects with sulphonylureas. In general, the IDD (type I) subjects were classified on the basis of documented evidence of being ketosis prone in the history and/or episodes of ketoacidosis and/or on the basis of weight and abruptness of onset of the diabetic symptoms. There were 39 men and 23 women. The IDD (type I) subjects were comparable by age and sex to the control subjects (table II). The NIDD (type II) subjects were comparable by sex to the other subjects, by blood sugar (enzymatic determination) and glycosylated hemoglobin (measured by a microcolumn technique (Bio Rad, Munich, FRG) to the IDD (type I) subjects.

Subjects continued to consume their usual diet during the study. After an overnight fast venous blood samples were drawn and centrifuged immediately. VLDL were isolated from serum by ultracentrifugation at 1.006 g/ml at 50,000 rpm for 22 h at 4 °C in Beckman 50.3 rotors, and the lipoprotein fractions were recovered by tube slicing [25]. HDL in the infranate were separated from LDL by precipitation of LDL using a combination of sodium phosphotungstate and magnesium chloride (Boehringer kit, Mannheim, FRG). The mass of cholesterol and triglycerides was determined enzymatically (Boehringer kits). Concentrations of apolipoproteins A-I, B, and E in serum were measured by endpoint immunonephelometry [26]. The coefficients of variation of these assays ranged from 2 to 5%. Statistical analysis ($p < 0.01$) was made using the Wilcoxon rank test for comparisons among the groups.

Results

Significant elevations of serum triglycerides and VLDL lipids were observed in the NIDD (type II) subjects versus control subjects (table III). Total cholesterol levels were elevated only in subjects without sulphonylureas. However, LDL cholesterol was higher in all groups of diabetics compared with controls. HDL cholesterol was significantly decreased in the NIDD (type II) subjects. In comparison with the IDD (type I) subjects the NIDD (type II) subjects had higher levels of VLDL lipids, LDL cholesterol (only type II without sulphonylurea), but lower levels of HDL cholesterol. The mean cholesterol/triglyceride mass ratio within VLDL was, furthermore, significantly elevated in the NIDD (type II) group compared with control and IDD (type I) subjects ($\pm$ SD): 0.33 ± 0.09 and 0.31 ± 0.13 vs. 0.26 ± 0.08 (controls) and 0.24 ± 0.07 (type I).

Apolipoprotein A-I levels were significantly higher in IDD (type I) subjects compared with controls and other diabetic groups. Apolipo-

Table I. Lipoprotein characteristics in diabetes mellitus

	IDD (type I) subjects	NIDD (type II) subjects
VLDL	VLDL cholesterol ↑ [4]	VLDL triglycerides ↑ [5, 6] VLDL Apo B ↑ [5, 7] VLDL Apo E ↑ [7] Serum Apo E ↑ [8]
LDL	No changes (+ Apo B) [9] LDL cholesterol ↑ [4]	No changes (+ Apo B) [9] LDL cholesterol (women) ↑ [10] LDL/HDL ratio ↑ [11]
HDL	HDL cholesterol ↓ [4, 9, 12] HDL cholesterol ↑ [13–16] Apo A-I ↓ [9, 12] Apo A-I ↑ [15–17] HDL_3 ↑ [18]	HDL cholesterol ↓ [9, 10, 12, 19, 20] Apo A-I no changes [19] Apo A-I ↓ [9, 12]

Table II. Characteristics of normal and diabetic subjects (mean ± SD)

	n	Age, years	Relative body weight, %[1]	10.00 h blood sugar, mg/dl	Glycosylated hemoglobin, %
Controls	13	39 ± 15	94 ± 9	75 ± 5	4.4 ± 1.2
IDD (type I) subjects	18	40 ± 16	95 ± 13	218 ± 60	7.5 ± 1.3
NIDD (type II) subjects without sulphonylurea	19	59 ± 14	100 ± 14	188 ± 42	6.9 ± 1.1
NIDD (type II) subjects with sulphonylurea	12	60 ± 8	101 ± 18	210 ± 55	7.3 ± 1.1

[1] 100% = height (cm) − 100.

protein B levels were higher in NIDD (type II) subjects, with sulphonylureas, while the apolipoprotein E concentrations in both groups of NIDD (type II) subjects significantly exceeded that of controls (+32.9% and +29.2%) and of IDD (type I) subjects (+25.4% and +22.6%), respectively. The atherogenic ratio LDL/HDL cholesterol

Table III. Serum and lipoprotein lipids and apolipoproteins in normals and diabetics

	Controls	IDD (type I) subjects	NIDD (type II) subjects without sulphonylurea	NIDD (type II) subjects with sulphonylurea
Cholesterol	161 ±31	206 ±45	248 ±60*	208 ± 56
Triglycerides	97 ±53	119 ±48	204 ±90*	190 ±102*
VLDL cholesterol	10 ± 6	16 ±11	34 ±20*,+	30 ± 23*,+
VLDL triglycerides	37 ±34	57 ±37	102 ±50*,+	96 ± 61*,+
LDL cholesterol	103 ±26	141 ±44*	180 ±49*,+	144 ± 45*
HDL cholesterol	42 ±16	48 ±16	32 ± 9*,+	34 ± 9*,+
Apolipoprotein A-I	140 ±12	172 ±34*	145 ±26+	158 ± 26+
Apolipoprotein B	73 ±13	81 ±23	106 ±21*,+	85 ± 27
Apolipoprotein E	8.5± 1.4	9.0± 2.0	11.3± 1.2*,+	11.0± 2.1*,+
LDL/HDL cholesterol	2.4± 0.9	3.1± 1.6*	5.6± 2.1*,+	3.9± 1.2*
Apolipoproteins B/A–I	0.5± 0.1	0.5± 0.2	0.7± 0.2*,+	0.6± 0.2

Mean ± SD (mg/dl). *,+ $p < 0.01$ (* diabetics vs. controls; + NIDD (type II) vs. IDD (type I)).

was significantly elevated in all diabetic groups. The ratio apolipoproteins B/A-I was observed to be increased only in the diabetic subjects without sulphonylureas.

Discussion

The relative weights of the NIDD (type II) subjects were greater than that of the IDD (type I) and normal subjects, as is usually seen in a group of NIDD patients [27]. In addition to mild obesity, patients with NIDD often have normal or even increased serum insulin concentrations [28]. This may explain the significant elevations of the VLDL fraction in NIDD subjects as compared with normal or IDD subjects [29]. The accompanying increase of the apolipoprotein E concentration in this study was greater than would be predicted on the basis of their mildly elevated triglyceride levels. This finding suggests that apolipoprotein E was a factor in abnormal cholesterol metabolism of NIDD subjects. The data of *Fielding* et al. [30] indicate an accumulation of apolipoprotein E-enriched VLDL with an increased free cholesterol content. NIDD (type II) subjects seem to have similar metabolic defects in the transfer of cholesteryl esters as dyslipoproteinemics or hyperbetalipoproteinemics with a high risk for atherosclerosis [31], leading to a disturbed cholesterol transport from peripheral cells to the liver.

The significant decrease of HDL cholesterol in NIDD subjects is in agreement with earlier reports (table I). One explanation is the known inverse relationship between HDL and serum and VLDL triglycerides [32]. However, a decrease of HDL due to insulin resistance cannot be ruled out, because sufficient insulinization results in a normalization of serum lipoprotein patterns [33, 34]. In normals, insulin administration also lowers the LDL/HDL cholesterol ratio [35]. Thus, higher apolipoprotein A-I levels in IDD subjects than in normals seem to be the result of an adequate control of the diabetes.

In summary, NIDD (type II) subjects are characterized by an increased concentration of apolipoprotein E-containing lipoproteins. Insulin treatment counteracts the lipoprotein disturbance in diabetes mellitus by increasing the number of HDL particles.

References

1 Bierman, E.L.; Brunzell, J.D.: Interrelation of atherosclerosis, abnormal lipid metabolism, and diabetes mellitus; in Katzen, Mahler, Advances in modern nutrition, vol. 7, pp. 187–210 (Wiley, New York 1978).
2 Sorge, F.; Schwartzkopff, W.; Neuhaus, G.A.: Insulin response to oral glucose in patients with a previous myocardial infarction and in patients with peripheral vascular disease: hyperinsulinemia and its relationship to hypertriglyceridemia and overweight. Diabetes 25: 586–594 (1976).
3 Beach, K.W.; Brunzell, J.D.; Conquest, L.L.; Strandness, D.E.: The correlation of arteriosclerosis obliterans with lipoproteins in insulin-dependent and non-insulin-dependent diabetes. Diabetes 28: 836–840 (1979).
4 Lopes-Virella, M.F.; Wohltmann, H.J.; Loadholt, G.B.; Buse, M.Y.: Plasma lipids and lipoproteins in young insulin-dependent diabetic patients: relationship with control. Diabetologia 21: 216–223 (1978).
5 Kissebah, A.H.; Alsfarsi, S.; Evans, D.J.; Adams, P.W.: Integrated regulation of very low-density lipoprotein triglyceride and apolipoprotein B kinetics in non-insulin-dependent diabetes mellitus. Diabetes 31: 217–225 (1982).
6 Howard, B.V.; Reitman, J.S.; Vasquez, B.; Zech, L.: Very low-density lipoprotein triglyceride metabolism in non-insulin-dependent diabetes mellitus. Diabetes 32: 271–276 (1983).

7   Weisweiler, P.; Drosner, M.; Schwandt, P.: Dietary effects on very low-density lipoproteins in type 2 (non-insulin dependent) diabetes mellitus. Diabetologia 23: 101–103 (1982).
8   Fielding, C.J.; Reaven, G.M.; Fielding, P.E.: Human non-insulin-dependent diabetes: identification of a defect in plasma cholesterol transport normalized in vivo by insulin and in vitro by selective immunoadsorption of apolipoprotein E. Proc. natn. Acad. Sci. USA 79: 6365–6369 (1982).
9   Schernthaner, G.; Kostner, G.M.; Dieplinger, H.; Prager, R.; Mühlhauser, J.: Apolipoproteins (A-I, A-II, B), Lp (a) lipoprotein and lecithin: cholesterol acyltransferase activity in diabetes mellitus. Atherosclerosis 49: 277–293 (1983).
10 Howard, B.V.; Knowler, W.C.; Vasques, B.; Kennedy, A.L.; Pettit, D.J.; Bennett, P.H.: Plasma and lipoprotein cholesterol and triglyceride in the Pima Indian population. Arteriosclerosis 4: 462–471 (1984).
11 Schmitt, J.K.; Poole, J.R.; Lewis, S.B.; Shore, V.G.; Maman, A.; Baer, R.M.; Forsham, P.H.: Hemoglobin A_1 correlates with the ratio of low-to-high density lipoprotein cholesterol in normal weight type II diabetics. Metabolism 31: 1084–1089 (1982).
12 Briones, E.R.; Mao, S.J.T.; Palumbo, P.J.; O'Fallon, W.M.; Chenoweth, W.; Kottke, B.A.: Analysis of plasma lipids and apolipoproteins in insulin-dependent and non-insulin-dependent diabetics. Metabolism 33: 42–49 (1984).
13 Klujber, L.; Molnar, D.; Kardos, M.; Jászai, V.; Soltész, G.; Mestyan, J.: Metabolic control, glycosylated haemoglobin, and high-density lipoprotein cholesterol in diabetic children. Eur. J. Pediat. 132: 289–297 (1979).
14 Nikkilä, E.A.: High-density lipoproteins in diabetes. Diabetes 30: 82–87 (1981).
15 Eckel, R.H.; Albers, J.J.; Cheung, M.C.; Wahl, P.W.; Lindgren, F.T.; Bierman, E.L.: High-density lipoprotein composition in insulin-dependent diabetes mellitus. Diabetes 30: 132–138 (1981).
16 Ewald, U.; Gustafson, S.; Tuvemo, T.; Vessby, B.: Increased high-density lipoproteins in diabetic children. Eur. J. Pediat. 142: 154–156 (1984).
17 Bachem, M.G.; Paschen, K.; Strobel, B.; Jastram, H.U.; Janssen, E.G.; Dati, F.: Beziehungen zwischen Lipidstoffwechsel und glycosylierten Hämoglobinen beim juvenilen Diabetes mellitus. Klin. Wschr. 60: 497–503 (1982).
18 Durrington, P.N.: Serum high-density lipoprotein cholesterol subfractions in type I (insulin-dependent) diabetes mellitus. Clinica chim. Acta 120: 21–28 (1982).
19 Taylor, K.G.; Wright, A.D.; Carter, T.J.N.; Valente, A.J.; Betts, S.A.; Matthews, K.A.: High-density lipoprotein cholesterol and apolipoprotein A-I levels at diagnosis in patients with non-insulin-dependent diabetes. Diabetologia 20: 535–539 (1981).
20 Biesbroeck, R.; Albers, J.J.; Wahl, P.; Weinberg, C.; Bassett, M.; Bierman, E.: Abnormal composition of high-density lipoproteins in non-insulin-dependent diabetics. Diabetes 31: 126–131 (1982).
21 Kannel, W.B.; Castelli, W.P.; Gordon, T.: Cholesterol in the prediction of atherosclerotic disease. Ann. intern. Med. 90: 85–91 (1979).
22 Tatami, R.; Mabuchi, H.; Ueda, K.; Ueda, R.; Haba, T.; Kametani, T.; Ito, S.; Koizumi, J.; Ohta, M.; Miyamoto, S.; Nakayama, A.; Kanaya, H.; Oiwake, H.: Intermediate-density lipoprotein and cholesterol-rich very low-density lipoprotein in angiographically determined coronary artery disease. Circulation 64: 1174–1184 (1981).

23 Castelli, W.P.; Doyle, J.T.; Gordon, T.; Hames, C.G.; Hjortland, M.C.; Hulley, S.B.; Kagan, A.; Zukel, W.J.: HDL cholesterol and other lipids in coronary heart disease. The cooperative lipoprotein phenotyping study. Circulation 55: 767–772 (1977).
24 Avogaro, P.; Cazzolato, G.; Bittolo Bon, G.; Quinci, G.B.: Are apolipoproteins better discriminators than lipids for atherosclerosis? Lancet i: 901–903 (1979).
25 Havel, R.J.; Eder, H.A.; Bragdon, J.H.: The distribution and chemical composition of ultracentrifugally separated lipoproteins in human serum. J. clin. Invest. 34: 1345–1353 (1955).
26 Weisweiler, P.; Schwandt, P.; Friedl, C.: Determination of human apolipoproteins A-I, B, and E by laser nephelometry. J. clin. Chem. clin. Biochem. 22: 113–118 (1984).
27 Bennet, P.H.; Rushforth, N.B.; Miller, M.; Le Compte, P.M.: Epidemiological studies of diabetes in the Pima Indians. Recent Prog. Horm. Res. 32: 333–376 (1976).
28 Reaven, G.M.; Greenfield, M.S.: Diabetic hypertriglyceridemia. Evidence for three clinical syndromes. Diabetes 30: 66–75 (1981).
29 Tobey, T.A.; Greenfield, M.; Kraemer, F.; Reaven, G.M.: Relationship between insulin resistance, insulin secretion, very low-density lipoprotein kinetics, and plasma triglyceride levels in normotriglyceridemic man. Metabolism 30: 165–171 (1981).
30 Fielding, C.J.; Reaven, G.M.; Lin, G.; Fielding, P.E.: Increased free cholesterol in plasma low- and very low-density lipoproteins in non-insulin-dependent diabetes mellitus: its role in the inhibition of cholesteryl ester transfer. Proc. natn. Acad. Sci. USA 81: 2512–2516 (1984).
31 Fielding, P.E.; Fielding, C.J.; Havel, R.J.; Kane, J.P.; Tun, P.: Cholesterol net transport, esterification, and transfer in human hyperlipidemic plasma. J. clin. Invest. 71: 449–460 (1983).
32 Kinnunen, P.K.J.: High-density lipoprotein may not be antiatherogenic after all. Lancet ii: 34–35 (1979).
33 Pietri, A.; Dunn, F.L.; Raskin, P.: The effect of improved diabetic control on plasma lipid and lipoprotein levels. Diabetes 29: 1001–1005 (1980).
34 Sosenko, J.M.; Breslow, J.L.; Miettinen, O.S.; Gabbay, K.H.: Hyperglycemia and plasma lipid levels. A prospective study of young insulin-dependent diabetic patients. New Engl. J. Med. 302: 650–654 (1980).
35 Sadur, C.N.; Eckel, R.H.: Insulin-mediated increases in the HDL cholesterol/cholesterol ratio in humans. Arteriosclerosis 3: 339–343 (1983).

Priv. Doz. Dr. med. P. Weisweiler, Second Medical Department,
Grosshadern Clinic, University of Munich,
Marchioninistrasse 15, D-8000 Munich 70 (FRG)

Apolipoprotein E Isoforms in Diabetes

Susan Black[a], R.V. Brunt[a], J.P.D. Reckless[a,b]

[a]Department of Biochemistry, University of Bath, and [b]Clinical Investigation Department, Royal United Hospital, Bath, England

Introduction

Apolipoprotein E (ApoE), a constituent of triglyceride-rich lipoproteins, mediates interactions with lipoprotein receptors [1] leading to lipoprotein uptake and degradation. ApoE occurs as a number of isoproteins which are unequivocably separable by two-dimensional electrophoresis [2]. There are three major isoforms $ApoE_4$, E_3 and E_2, which differ by single amino acid substitutions involving cysteine-arginine interchange, containing 0, 1 and 2 cysteine residues/mol respectively [3]. The resulting charge differences ensure separation by isoelectric focusing (IEF). Studies of ApoE pattern inheritance [4] indicate a single gene locus with 3 alleles e^2, e^3 and e^4 specifying respectively E_2, E_3 and E_4, and giving 6 phenotypes distinguished by IEF [5, 6]. Minor isoforms arise by differing degrees of sialylation of major isoforms [4], leading to a charge shift similar to that from cysteine substitution. Thus, monosialylated $ApoE_3$ ($ApoE_3S_1$) co-focuses with $ApoE_2$, while $ApoE_1$ and $ApoE_1'$ species are more sialylated forms of $ApoE_2$, $ApoE_3$ and $ApoE_4$. $ApoE_2$ has markedly decreased binding to ApoB/E receptors compared to $ApoE_3$ and $ApoE_4$ [7]. Sialylated forms have received little attention, although perhaps increased in diabetes [8], and this investigation was therefore undertaken.

Methods

Very low density lipoprotein (VLDL) aproproteins were prepared by ultracentrifugation, delipidation and resolubilization [6]. ApoVLDL (50 µg/gel) were electro-

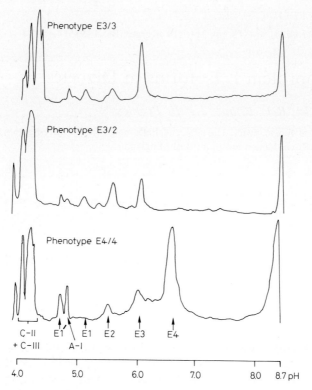

Fig. 1. IEF gels of VLDL apolipoproteins.

phoresed in 12.5% polyacrylamide tube gels with 0.1% sodium dodecyl sulphate, stained with Coomassie Brilliant Blue R-250, destained and scanned, and proportion of ApoE determined by peak area integration.

IEF gels [6] contained 4% pH 5–7 and 1% pH 3–10 ampholines (Bio-Rad). 50 μg ApoVLDL were loaded onto prefocused gels and focused at 500 V for 16 h at 18 °C.

Results

ApoE as a proportion of ApoVLDL was variable, with some increase in diabetics at the expense of ApoC (table I). Figure 1 shows typical IEF scans for 3 phenotypes. The $E_{3/3}$ phenotype is of interest, for in the absence of true $ApoE_2$ and $ApoE_4$ the $ApoE_2$ position represents $ApoE_3S_1$. In the $E_{3/2}$ phenotype the $ApoE_2$ position is occupied both by $ApoE_2$ and $ApoE_3S_1$.

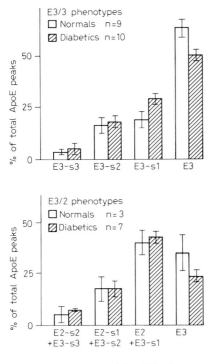

Fig. 2. Percent area of ApoE isoforms on IEF gels.

Table I. Comparison of VLDL apolipoprotein composition by SDS-Page

Apolipoprotein	n	Percent area of densitometric scan (mean ± SE)			
		E	A-I	A-IV	C-I, C-II, C-III, A-II
Normals	7	12 ± 2	3 ± 1	3 ± 2	82 ± 3
Diabetics	20	21 ± 11	4 ± 3	4 ± 3	71 ± 15

Isoform distribution in $ApoE_{3/3}$ phenotypes (fig. 2) showed significantly decreased $ApoE_3$ in diabetics compared to normals with compensatory increase in sialylated isoforms, largely of $ApoE_3S_1$. $ApoE_{3/2}$ phenotypes showed similar trends with lower $ApoE_3$ in diabetics, although the number of normals is inadequate.

Discussion

In ApoE$_{3/3}$ phenotypes the ApoE$_2$ position has been assumed to be occupied by ApoE$_3$S$_1$. However, if ApoE was glycated at lysyl residues the isoform mobility on IEF could be similar to monosialylation. Such a distinction could be made by neuraminidase treatment of the isoforms, and is underway.

References

1. Mahley, R.W.: Atherogenic hyperlipoproteinaemia. The cellular and molecular biology of plasma lipoproteins altered by dietary fat and cholesterol. Med. Clins N. Am. *66:* 375–402 (1982).
2. Zannis, V.I.; Breslow, J.L.: Characterisation of a unique human apolipoprotein variant associated with type III hyperlipoproteinaemia. J. biol. Chem. *255:* 1759–1762 (1980).
3. Weisgraber, K.H.; Rall, S.C.; Mahley, R.W.: Human E apoprotein heterogeneity. Cysteine-arginine interchanges in the amino acid sequence of the ApoE isoforms. J. biol. Chem. *256:* 9077–9083 (1981).
4. Zannis, V.I.; Breslow, J.L.: Human very low density lipoprotein apolipoprotein E isoprotein polymorphism is explained by genetic variation and post-translational modification. Biochemistry *20:* 1033–1041 (1981).
5. Utermann, G.; Steinmetz, A.; Weber W.: Genetic control of human apolipoprotein E polymorphism. Comparison of one- and two-dimensional techniques of isoprotein analysis. Hum. Genet. *60.* 344–351 (1982).
6. Warnick, G.R.; Mayfield, C.; Albers, J.J.; Hazzard, W.R.: Gel isoelectric focussing method for specific diagnosis of familial hyperlipoproteinaemia type III. Clin. Chem. *25:* 279–284 (1979).
7. Weisgraber, K.H.; Innerarity, T.L.; Mahley R.W.: Abnormal lipoprotein receptor-binding activity of the human E apoprotein due to cysteine-arginine interchange at a single site. J. biol. Chem. *257:* 2518–2523 (1982).
8. Weisweiler, P.; Jungst, D.; Schwandt, P.: Quantitation of apolipoprotein E isoforms in diabetes mellitus. Hormone metabol. Res. *15:* 201–205 (1983).

Dr. J. Reckless, Royal United Hospital, Bath BA1 3NG (England)

Serum HDL Concentrations in Patients with Type I and Type II Diabetes mellitus

R. Carmena, J.F. Ascaso, S. Serrano, J. Martinez-Valls, P. Soriano

Department of Medicine, Facultad de Medicina, Hospital Clínico Universitario, Valencia, Spain

Numerous clinical and epidemiological studies have confirmed the existence of an inverse and independent relationship between coronary heart disease (CHD) and serum high density lipoprotein (HDL) cholesterol concentration [1].

Accelerated atherosclerosis and increased risk for CHD are major complications of both insulin-dependent (type I) and noninsulin-dependent (type II) diabetes mellitus [2]. The factors predisposing diabetic patients to early development of atherosclerosis are not well understood but lipid disorders and abnormal serum lipoprotein concentrations are frequently found among diabetics. Of particular interest, HDL cholesterol (HDL-C) concentration has been shown to be reduced in some but not all groups of diabetic patients [3–8]. The discrepancies observed can probably be explained by the highly heterogenous nature of diabetes mellitus and the fact that some of the published studies used mixed populations of men and women with type I and type II diabetes treated in many different fashions.

Methods

We have studied the serum HDL-C levels in different groups of male and female patients with type I and type II diabetes mellitus both at the time of diagnosis and after different follow-up periods of treatment with diet and intermediate-acting insulin twice daily or diet and glibenclamide 5–15 mg/day. HDL-C was measured by the method of *Burstein* et al. [9].

Results

Table I includes data on 17 male patients (mean age 20.1 ± 5.0 years) with insulin-dependent diabetes of at least 1 year duration. Up to the time of the study (designated 'basal' in table I) they had received a single morning injection of intermediate-acting insulin and their diabetic control was only moderately acceptable, as expressed by a mean glycosylated hemoglobin value of 14.1% (normal in our laboratory: 5.5–8.5%). However, the HDL-C concentration was within normal limits: 46.7 vs. 48.3 mg/dl in our control group of normal men. All patients were instructed in diabetic autocontrol, switched to a twice daily intermediate insulin regimen and followed for a 6-month period (designated 'final' in table I). A better diabetic control was thus achieved, as expressed by lower mean fasting blood sugar and glycosylated hemoglobin values. The HDL-C concentration rose to 58.7 mg/dl, well above the values observed in normal men.

Table II shows the mean serum lipid levels in obese and nonobese male patients with recently diagnosed, untreated, type II diabetes mellitus. Compared with a control group of normal men both diabetic groups had significantly lower levels of serum HDL-C and higher serum triglyceride values.

Table III shows mean serum lipid values in obese women with type II diabetes and in two groups of obese, nondiabetic women separated according to their serum triglyceride levels and previously studied by us [10]. Prior to starting therapy with hypocaloric diet and, in the diabetic group, 5–15 mg/day of glibenclamide, the three groups showed (designated 'A' in table III) values of HDL-C significantly lower than in the control group. After a 10-month period of treatment all groups had lost weight and the obese nondiabetics had elevated their HDL-C to normal levels. However, in the diabetic group the HDL-C concentration remained unchanged.

Discussion

In our group of 17 young males with type I diabetes the mean HDL-C value after at least 1 year of treatment with a single injection of intermediate-acting insulin was not different from the value observed in a control group of normal men, 46.7 ± 7.6 and 48.3 ± 6.6 mg/dl,

Table I. 17 insulin-dependent (type I) diabetic males switched from one to two daily insulin injections and followed during 6 months (mean age: 20.1 ± 5.0 years)

		FBS mg/dl	HbA$_1$ %	T chol mg/dl	TG mg/dl	HDL-C mg/dl	LDL-C mg/dl	VLDL-C mg/dl
Basal	x̄	208.3	14.1	199.2	285.1	46.7	108.0	55.2
	±SD	52.1	4.9	58.7	79.0	7.6	43.8	18.2
Final	x̄	150.3	8.0	175.6	217.1	58.7	71.0	45.0
	±SD	50.9	1.1	38.9	64.5	9.6	34.6	13.2
	t	4.58	4.38	1.84	2.22	−3.80	2.13	2.13
	p	<0.001	<0.001	NS	<0.05	<0.01	<0.05	<0.05

Table II. Serum lipid values at the time of diagnosis of type II diabetes mellitus in male patients and normal controls

		Age years	TG mg %	T chol mg %	HDL-C mg %
Type II DM					
Obese	x̄	53.1	233.4	242.2	34.9
n = 15	±SD	12.0	68.4	52.0	5.2
Nonobese	x̄	49.1	219.5	250.0	35.5
n = 14	±SD	13.7	65.3	54.6	8.5
Controls	x̄	46.6	113.9	176.2	48.3
n = 21	±SD	10.0	22.5	30.1	6.6
			p < 0.01	p < 0.01	p < 0.01

respectively. We have thus confirmed that type I diabetic patients treated with insulin and kept under fair metabolic control (Hemoglobin A$_1$ ≤14%) show normal HDL-C values, as previously reported by others [11–13]. Following the institution of a tighter diabetic control with two daily injections of intermediate-acting insulin, which meant a net increase in the total daily dose of each patient of 5–15 units, the level of glycosylated hemoglobin dropped to 8% and the HDL-C rose to 58.7 ± 9.6 mg/dl, significantly above the values observed in normal men of comparable age. Similar results have been published by others [11, 13]. Improved diabetic control could be one explanation considered by some authors [12]. On the other hand, *Nikkilä* [7, 11] has shown that

Table III. Serum lipid values in obese type II diabetic females and obese nondiabetic females before (A) and 10 months after (B) treatment with oral hypoglycemic agents and hypocaloric diet

		RBW %		TG mg/dl		T chol mg/dl		HDL-C mg/dl		FBS mg/dl	HbA %
Diabetics n=22, age x̄=40±10 years	A	176.0	30.5	239.6	76.5	254.1	59.8	34.0*	10.1	221.2±43.0	18.0±4.1
	B	156.2	26.7	184.0	64.0	223.6	48.0	36.1*	8.9	169.0±51.5	10.5±2.0
Non-diabetics Hyper-TG n=20, age x̄=38±12 years	A	186.7	34.4	225.3	48.6	221.8	54.0	41.5*	9.6		
	B	160.0	33.3	155.3	28.0	211.3	37.7	48.2	11.1		
Normo-TG n=20, age x̄=36±11 years	A	174.8	20.5	148.5	15.5	185.0	41.6	42.5*	10.4		
	B	146.2	16.0	135.0	28.8	191.0	48.2	53.6	11.2		
Controls n=15 females, age x̄=35±9 years		—		121.3	22.7	172.8	20.5	54.3	10.0		

Values are x̄±SD. * p<0.01 vs. controls.

high free insulin levels, present in the plasma of insulin-treated diabetic patients, increase the postheparin plasma lipoprotein lipase (LPL) activity which in turn increases VLDL catabolism. The same author has demonstrated a significant correlation between HDL-C concentration and plasma LPL activity in diabetics.

We have also studied two different groups of male and female patients with type II (noninsulin-dependent) diabetes mellitus. They had slightly elevated serum triglyceride levels and their HDL-C concentration at the time of diagnosis was significantly lower than in the control groups, a finding reported also by *Taylor* et al. [14]. The fact that HDL-C levels were similar in men and women suggests that the presence of diabetes causes a greater reduction in HDL in women than in men. Following therapy with diet and glibenclamide we failed to observe significant changes in HDL-C levels, despite improved metabolic control. This confirms reports by other authors [15, 16] and probably reflects some degree of insulin resistance and low plasma LPL activity more than a direct effect of oral hypoglycemic agents on HDL metabolism.

References

1 Gordon, T.; Kannel, W.B.; Castelli, W.P.; Dawbert, T.R.: Lipoproteins, cardiovascular disease and death. The Framingham Study. Archs intern. Med. *141:* 1128–1131 (1981).
2 Kannel, W.B.; McGee, D.L.: Diabetes and cardiovascular disease. The Framingham Study. J. Am. med. Ass. *241:* 2035–2038 (1979).
3 Barr, D.P.; Russ, E.M.; Eder, H.A.: Protein-lipid relationships in human plasma. 2. Atherosclerosis and related conditions. Am. J. Med. *11:* 480–483 (1951).
4 Lopes-Virella, M.F.L.; Stone, P.G.; Colwell, J.A.: Serum high density lipoprotein in diabetic subjects. Diabetologia *13:* 285–291 (1977).
5 Bar-On, H.; Landau, D.; Berry, E.: Serum high-density lipoprotein and University Group Diabetes Program results. Lancet *i:* 761 (1977).
6 Howard, B.V.; Savage, P.J.; Bennion, L.J.; Bennett, P.H.: Lipoprotein composition in diabetes mellitus. Atherosclerosis *30:* 153–162 (1978).
7 Nikkilä, E.A.: Metabolic and endocrine control of plasma high density lipoprotein concentration; in Grotto, Miller, Oliver, High density lipoproteins and atherosclerosis, pp. 177–192 (Elsevier, Amsterdam 1978).
8 Pietri, A.; Dunn, F.L.; Raskin, P.: The effect of improved diabetic control on plasma lipid and lipoprotein levels. Diabetes *29:* 1001–1005 (1980).
9 Burstein, M.; Scholnick, H.R.; Morfin, R.: Rapid method for the isolation of lipoproteins from human serum by precipitation with polyanions. J. Lipid Res. *11:* 583–595 (1970).

10   Carmena, R.; Ascaso, J.F.; Tebar, J.; Soriano, J.: Changes in plasma high-density lipoproteins after body weight reduction in obese women. Int. J. Obes. 8: 135–140 (1984).
11   Nikkilä, E.A.; Hormila, P.: Serum lipids and lipoproteins in insulin-treated diabetes: demonstration of increased high density lipoprotein concentrations. Diabetes 27: 1078–1086 (1978).
12   Sosenko, J.M.; Breslow, J.L.; Miettinen, O.S.; Gabbay, K.H.: Hyperglycemia and plasma lipid levels. A prospective study of young insulin-dependent diabetic patients. New Engl. J. Med. 302: 650–654 (1980).
13   Falko, J.M.; O'Dorisio, T.M.; Catalands, S.: Improvement of high-density lipoprotein cholesterol levels. J. Am. med. Ass. 247: 37–39 (1982).
14   Taylor, K.G.; Wright, A.D.; Carter, T.J.N.; Valente, A.J.; Betts, S.A.; Matthews, K.A.: High-density lipoprotein cholesterol and apolipoprotein A-I levels at diagnosis in patients with non-insulin dependent diabetes. Diabetologia 20: 535–539 (1981).
15   Eder, H.A.; Gidez, L.I.: High density lipoproteins, HDL_2 and HDL_3, in healthy subjects and in patients with coronary heart disease and with diabetes; in Noseda, Fragiacomo, Fumagalli, Paoletti, Lipoproteins and coronary atherosclerosis, pp. 45–52 (Elsevier, Amsterdam 1982).
16   Huupponen, R.K.; Viikari, J.S.; Saarimaa, H.: Correlation of serum lipids with diabetes control in sulfonylurea-treated diabetic patients. Diabetes Care 7: 575–578 (1984).

R. Carmena, MD, Department of Medicine, Facultad de Medicina,
Hospital Clinico Universitario, E-46010 Valencia (Spain)

The Association between Primary Gout and Hypertriglyceridaemia May Be due to Genetic Linkage

G.A.A. Ferns[a], J. Lanham[b], D.J. Galton[a]

Departments of [a]Lipid Research and [b]Rheumatology, St. Bartholomew's Hospital, London, UK

Introduction

Primary gout is a rare metabolic disorder of urate metabolism, characterized by urate crystal arthropathy and hyperuricaemia (a serum urate level >0.42 mmol/l). It is of particular interest because it associates with premature atherosclerosis and hypertriglyceridaemia (a serum triglyceride >2.0 mmol/l). A feature of gouty hypertriglyceridaemia is its persistence despite drug therapy that causes abatement of symptoms and normouricaemia. For this reason it was considered possible that gout and hypertriglyceridaemia are genetically linked disorders, with genes influencing tissue levels of uric acid and serum triglyceride being closely located on the same chromosome. To investigate this hypothesis we have determined the distribution of allelic variants at the apoprotein AI-CIII gene cluster locus (localized to the long arm of chromosome 11) in a group of subjects with gout. The common allele at this locus in Caucasians in the S_1 allele. When genomic DNA is digested with the restriction enzyme SstI, the S_1 allele is characterized by being cleaved into two ApoAI gene-related fragments of 4.2 and 5.7 kilo base pairs (kbp) respectively. An uncommon S_2 allele at this locus is found in 30% of Caucasian hypertriglyceridaemics but only 5% of normolipidaemic Caucasian controls [1]. It differs from the S_1 allele in having a base transversion in the 3' noncoding region of the CIII gene; this creates an additional cleavage site for SstI and is thereby characterized by ApoAI gene-related fragments of 3.2 and 5.7 kbp respectively. DNA from heterozygous individuals with an S_1S_2 genotype would therefore produce ApoAI gene-related fragments of 3.2, 4.2 and 5.7 kbp, when digested with SstI.

Table I. Genotypic frequencies of a polymorphism in the 3' region of the apolipoprotein CIII gene in subjects with primary gout

	Gout patients	Normouricaemic controls	Normolipidaemic controls
Number of patients	30	41	33
Mean triglyceride mmol/l	2.75±1.927	1.52±0.761	1.23±0.361
p value		<0.01*	<0.001*
Mean BMI, kg/m^2	26.9±4.25	24.8±2.90	24.3±2.63
p value		<0.05*	<0.01*
Mean age, years	57.6±11.56	52.4±13.81	51.9±14.29
p value		>0.05*	>0.05*
Genotype S_1S_1	24 (80%)	40 (97%)	33 (100%)
Genotype S_1S_2	6 (20%)	1 (3%)	0 (0%)
Fisher's exact test p value		0.039	0.017

Comparison of genotypic frequencies of the polymorphic ApoAI-CIII gene cluster locus in subjects with primary gout, normouricaemic controls and normolipidaemic-normouricaemic controls. Body mass index was calculated as weight (kg)/height^2 (m^2). * p was calculated using an unpaired t test.

Subjects

Caucasian subjects with primary gout were recruited from our Rheumatology Clinic. A diagnosis was made using the following criteria: (a) four or more attacks of acute arthritis with at least one being an attack of podagra, in association with hyperuricaemia, or (b) acute arthritis with urate crystal arthropathy demonstrated by polarized light microscopy. Renal and hepatic function were assessed to exclude secondary causes of gout and the subjects were rendered asymptomatic by drug therapy. Controls were obtained from a health screening centre. Serum uric acid, fasting serum triglycerides and a fasting blood glucose were estimated for each individual. One subject with gout was also diabetic.

Methods

Leucocyte DNA was obtained from 10 ml whole blood anticoagulated with EDTA, using the method of *Kunkel* et al. [2]. Eight micrograms of DNA was digested with SstI, the restricted fragments separated by electrophoresis on a 1% agarose gel, Southern blotted onto a nitrocellulose filter, baked at 80 °C and then hybridized to an ApoAI radiolabelled gene-specific probe. Filters were subsequently washed and the hybridization bands visualized by autoradiography at $-70\,°C$. Subjects were genotyped S_1S_1 or S_1S_2 and the genotypic frequencies compared using Fisher's exact test.

Results and Discussion

Fifty percent of the gout group were hypertriglyceridaemic compared to less than 20% of normouricaemic controls. Those in the gout group were slightly older and heavier and this may account for part of the difference. However, we found that a significantly higher proportion of gout subjects were of an S_1S_2 genotype, and this was especially striking when compared with individuals who were both normouricaemic and normolipidaemic (table I). These findings suggest that gout and hypertriglyceridaemia may be related by the close proximity of disease-specific alleles that determine urate crystal deposition and affect triglyceride metabolism.

References

1 Rees, A.; Shoulders, C.; Stocks, J.; Baralle, F.; Galton, D.: DNA polymorphism adjacent to the human apolipoprotein AI gene: relation to hypertriglyceridaemia. Lancet *ii:* 444–446 (1983).
2 Kunkel, L.; Smith, K.; Boyer, S.; Borgaonkar, D.; Wachtel, S.; Muller, O.; Breg, W.; Jones, H.; Rary, J.: Analysis of human Y chromosome-specific reiterated DNA in chromosome variants. Proc. natn. Acad. Sci. USA *74:* 1245–1249 (1977).

Dr. G.A.A. Ferns, Department of Lipid Research, St. Bartholomew's Hospital, London EC1A 7BE (UK)

Lipoproteins and Lipoprotein Lipase Activities in Obese Type 2 Diabetics: Studies of the Relationship between Low Density Lipoproteins and the Skeletal Muscle Lipoprotein Lipase

B. Vessby, H. Lithell

Department of Geriatrics, Uppsala University, Uppsala, Sweden

Hyperlipidaemia is a common feature of diabetes mellitus type 2 [1], mainly due to increased very low density lipoprotein (VLDL) lipid levels. Increased concentrations of triglyceride (TG)-rich lipoproteins may be caused by increased production or impaired catabolism. After hydrolysis of the major part of the TG moiety by lipoprotein lipase the VLDL particles may be converted to low density lipoproteins (LDL). Thus the concentration of LDL is dependent on the rate of synthesis of VLDL as well as of the efficiency of the conversion of VLDL to LDL [2]. A low concentration of LDL may be due to either an increased rate of catabolism of LDL or to low VLDL synthesis and/or decreased lipoprotein lipase activity (LPLA).

We have investigated the relationships between the LPLA in adipose tissue (AT-LPLA) and skeletal muscle tissue (SM-LPLA) and the lipoprotein concentrations in obese diabetic subjects, both before and after a period of supplemented fasting [3]. The aim of the present report is to discuss the possible role of SM-LPLA as a determinant of the serum LDL concentrations.

Materials and Methods

Twenty patients with type 2 diabetes, 8 women and 12 men, were studied. All patients had a relative body weight above 1.19, according to Broca's index. The clinical characteristics of the patients have been given earlier [4]. The first 2 days after admission, the patients were given an ordinary isocaloric diabetic diet. The fasting period lasted for 3 weeks. All antidiabetic drugs were removed during fasting. During this period the patients consumed mainly juices from fruits and berries and vegetable beverages supplemented with 30 g of a nutrition formula preparation (Meritene; Wander AG, Bern, Switzerland). The approximate calorie intake was 200 kcal/day (0.9 MJ) [3].

Blood was drawn for analysis during the first 2 days after hospital admission and during the last 2 days of the fasting period. All tests were performed after an overnight fast. TG and cholesterol (Chol) concentrations in serum and in the isolated lipoprotein fractions were determined by using a combination of preparative ultracentrifugation [5] and heparin-manganese precipitation [6]. Specimens of abdominal adipose tissue and of skeletal muscle from the lateral vastus muscle were taken for determination of LPLA [7, 8]. The intravenous fat tolerance test (IVFTT) was performed as described by Carlson and Rössner [9].

Results

During fasting the body weight decreased from 102 ± 3.5 ($\bar{x} \pm$ SEM) to 91.7 ± 2.4 kg and the blood glucose was normalized. The lipoprotein lipid levels on admission and during the last day of the fasting period are shown in table I. The corresponding values for the fractional removal rate (K_2) of IVFTT, SM-LPLA and AT-LPLA are also given.

On admission there were significant negative correlations between K_2IVFTT and TG concentrations in VLDL, HDL and serum as well as a significant positive correlation between K_2IVFTT and SM-LPLA.

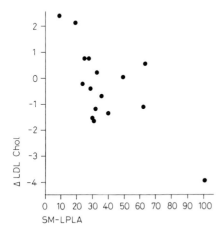

Fig. 1. Relationship between changes of LDL Chol during fasting (mmol/l) and the skeletal muscle lipoprotein lipase activity (SM-LPLA, mU/g) on admission.

Table 1. Lipoprotein lipid concentrations (mmol/l), lipoprotein lipase activity (mU/g) in adipose tissue (AT-LPLA) and skeletal muscle (SM-LPLA) and fractional removal rate (K_2, %/min) at the intravenous fat tolerance test (IVFTT) before and during fasting in obese type 2 diabetics (n = 20, $\bar{x} \pm$ SEM)

	VLDL		LDL		HDL		Serum		AT-LPLA	SM-LPLA	K_2IVFTT
	TG	Chol	TG	Chol	TG	Chol	TG	Chol			
On admission	3.9 ±0.9	1.6 ±0.4	0.6 ±0.03	4.0 ±0.4	0.29 ±0.02	0.91 ±0.04	4.9 ±0.9	6.8 ±0.4	159 ±13	38 ±5	2.7 ±0.2
During fasting	0.9*** ±0.08	0.4*** ±0.04	0.6 n.s. ±0.03	3.6 n.s. ±0.3	0.19** ±0.01	0.78** ±0.03	1.7*** ±0.1	4.9*** ±0.3	101*** ±9	19** ±2	3.7 n.s. ±0.4

** = $p < 0.01$ and *** = $p < 0.001$ compared with on admission; n.s. = not significant.

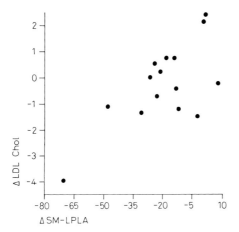

Fig. 2. Relationship between the changes of LDL Chol (mmol/l) and of SM-LPLA (mU/g) during fasting.

No similar relationship existed between K_2IVFTT and AT-LPLA or between AT-LPLA and the lipoprotein lipid concentrations. K_2IVFTT was significantly positively correlated with the LDL Chol concentrations. During fasting the reductions of the VLDL TG concentrations were most pronounced in subjects with high pretreatment values and with low K_2IVFTT.

The mean LDL Chol concentration on admission was normal, but the individual values showed a wide range from 2.08 to 9.82 mmol/l. Although the mean value of LDL Chol did not change during fasting, the individual changes were inversely correlated with the pretreatment SM-LPLA (r = 0.70, p < 0.01; fig. 1). Patients showing increasing LDL Chol concentrations during fasting apparently had low SM-LPLA on admission, while reductions took place in those with lower TG, higher LDL Chol and more efficient TG removal capacity. Compatible with this, there was a positive correlation between the individual changes of LDL Chol and those of SM-LPLA during fasting (r = 0.69, p < 0.01; fig. 2). Also, the changes of SM-LPLA and K_2IVFTT were significantly positively correlated (r = 0.55, p < 0.05).

Discussion

The positive correlation between LDL Chol and K_2IVFTT on admission indicated that the patients with the best capacity to eliminate circulating TG also had the highest LDL levels. The individual changes of LDL Chol during fasting were negatively correlated with the initial SM-LPLA, and the changes of LDL Chol and SM-LPLA were correlated. This is compatible with that the capacity to convert TG-rich lipoproteins to LDL is directly influenced by the activity of SM-LPLA. A high SM-LPLA allows an efficient conversion of VLDL to LDL, which subsequently increases. A very low SM-LPLA will on the other hand restrict the conversion resulting in a low LDL Chol level. Apparently, AT-LPLA does not influence this process, at least not in obese diabetic patients.

AT-LPLA and SM-LPLA have different K_m values for TG substrates [10]. This may be of physiological importance making SM-LPLA more active against the smaller, less TG-rich VLDL particles than AT-LPLA. It has been shown by *Clegg* et al. [11] that small VLDL particles are preferentially converted to LDL in contrast to big, TG-rich VLDL particles which do not undergo this transformation.

SM-LPLA is increased by clofibrate treatment [12] which is not infrequently associated with increasing LDL Chol in patients with low LDL Chol. After cessation of gemfibrozil treatment the TG levels increase, possibly mainly due to a reduced SM-LPLA [13]. The changes of LDL Chol are significantly correlated with those of SM-LPLA. Also, in hypertriglyceridaemic patients with severe coronary heart disease we have earlier shown that patients with low LDL Chol have lower SM-LPLA than those with high LDL Chol [14]. No similar correlation was seen in these patients to AT-LPLA.

In summary, we would like to suggest that the present data, supported by earlier studies, indicate that SM-LPLA may play an important role in the conversion of VLDL to LDL and thus be one determinant of the serum LDL concentrations.

References

1 Walden, C.; Knopp, R.; Wahl, P.; Beach, K.; Strandness, E.: Sex differences in the effect of diabetes mellitus on lipoprotein triglyceride and cholesterol concentrations. New Engl. J. Med. *311:* 953–959 (1984).

2   Vessby, B.; Lithell, H.: Dietary effect on lipoprotein levels in hyperlipoproteinemias. Determination of two subgroups of endogenous hypertriglyceridemia. Artery *1:* 63–85 (1974).
3   Vessby, B.; Selinus, I.; Lithell, H.: Serum lipoproteins and lipoprotein lipase activities in overweight, type II diabetics during and after supplemented fasting. Arteriosclerosis *5:* 93–100 (1985).
4   Vessby, B.; Boberg, M.; Karlström, B.; Lithell, H.; Werner, I.: Improved metabolic control after supplemented fasting in overweight type II diabetic patients. Acta med. scand. *216:* 67–74 (1984).
5   Havel, R.; Eder, H.; Bragdon, J.: The determination and chemical composition of ultracentrifugally separated lipoproteins in human serum. J. clin. Invest. *34:* 1345–1353 (1955).
6   Burstein, M.; Samaille, J.: Sur un dosage rapide du cholésterol lié aux alfa et aux béta lipoprotéins du sérum. Clinica chim. Acta *5:* 609 (1960).
7   Lithell, H.; Boberg, J.: A method of determining lipoprotein lipase activity in human adipose tissue. Scand. J. clin. Lab. Invest. *37:* 551–561 (1977).
8   Lithell, H.; Boberg, J.: Determinations of lipoprotein lipase activity in human skeletal muscle tissue. Biochim. biophys. Acta *528:* 58–68 (1978).
9   Carlson, L.; Rössner, S.: A methodological study of an intravenous fat tolerance test with intralipid emulsion. Scand. J. clin. Lab. Invest. *29:* 271–280 (1972).
10 Lithell, H.: Lipoprotein lipase activity in human skeletal-muscle and adipose tissue. Methodological and clinical studies; thesis. Acta Univ. Upsal. 272 (1977).
11 Clegg, R.; Munro, A.; Shepherd, J.; Packard, C.: Metabolic channelling of apolipoprotein B (apo B) in very low density lipoproteins (VLDL) (Abstract). Circulation *70:* II-118 (1984).
12 Lithell, H.; Boberg, J.; Hellsing, K.; Lundqvist, G.; Vessby, B.: Increase of the lipoprotein-lipase activity in human skeletal muscle during clofibrate administration. Eur. J. clin. Invest. *8:* 67–74 (1978).
13 Vessby, B.; Boberg, M.; Lithell, H.: The influence of gemfibrozil on lipoprotein composition, triglyceride removal capacity and fatty acid composition of the plasma lipid esters. Proc. R. Soc. Med., suppl. (in press, 1985).
14 Lithell, H.; Vessby, B.: Hypertriglyceridaemic CHD patients with low LDL-cholesterol levels have low skeletal muscle lipoprotein lipase activity. Abstr. 4th Int. Symp. on Atherosclerosis, West Berlin, 1982, p. 565.

B. Vessby, MD, Department of Geriatrics, Uppsala University, POB 12042, S-75012 Uppsala (Sweden)

Glycation of Very Low Density Lipoproteins in Diabetes

J.P.D. Reckless[a,b], Susan Black[b], R.V. Brunt[b]

[a]Clinical Investigation Department, Royal United Hospital, Bath, and
[b]Department of Biochemistry, University of Bath, England

Introduction

Low density lipoproteins (LDL), partly glycated in vitro, are incorporated into, and degraded by, human fibroblasts more slowly than control LDL not glycated in vitro [1], and at higher levels of glycation degradation may be completely inhibited. The effect could have physiological significance since such LDL gave a reduced fractional catabolic rate when injected into guinea pigs.

Existence of glycated LDL in human plasma was shown by measuring tritium incorporation into lysine-glucose bonds when LDL was reduced with sodium [^{3}H]borohydride (NaB[^{3}H]$_4$) (fig. 1). A 2- to 3-fold increase in LDL glycation was found in diabetes, and subsequently monoclonal antibodies specific to glucitollysine and mannitollysine confirmed this [2].

Since apolipoprotein E (ApoE)-containing lipoproteins react with similar receptors to the ApoB-containing LDL [3] very low density lipoproteins (VLDL) from diabetics might be expected to show increased glycation and reduced cellular uptake compared to control VLDL. Levels of glycation in VLDL from both controlled and ketoacidotic diabetics were compared with those in VLDL from normal subjects.

Methods

Ketoamine and hemiketal derivatives formed by glucose addition to the e-amino group of lysine are reduced by NaB[^{3}H]$_4$ to yield tritiated products (fig. 1). NaB[^{3}H]$_4$ was diluted with NaBH$_4$ made up in 0.5 mol/l NaOH and stored in small aliquots in liquid

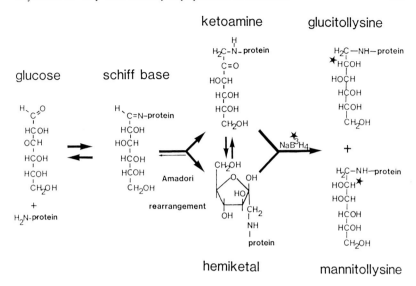

Fig. 1. Reaction scheme for the nonenzymic glycation of proteins.

nitrogen to minimize decomposition. Spontaneous hydrolysis of $NaB[^3H]_4$ to $[^3H]_2O$ on storage was estimated by reduction of benzaldehyde by $NaB[^3H]_4$, with extraction of the products into ethyl acetate. The aliquot was rejected if significant counts were found in the aqueous phase.

50 µg VLDL protein was reduced with 0.5 mmol/l $NaB[^3H]_4$ for 1 h on ice. 1 ml of 10% trichloracetic acid (TCA) was added. The protein pellet from centrifugation was dissolved in 100 µl 2 mol/l NaOH, the washing procedure being repeated another three times. Tritium content was estimated by scintillation counting with quench correction.

Results and Discussion

Preliminary experiments tested the method using bovine serum albumin (BSA) as a model protein (fig. 2). 10 mg BSA was incubated in phosphate-buffered saline pH 7.4 with 80 mmol/l glucose at 37 °C. Samples were removed at intervals, dialyzed and measured for borohydride reducible groups, which increased over a period of 13 days. Parallel experiments using [^{14}C]glucose showed a steady increase in labelled protein over 13 days. Incorporation into VLDL rose from 1.4 to 4.2 nmol H^-/mg VLDL protein over 13 days. Available lysyl residues may be lower in VLDL [4] compared to BSA [5].

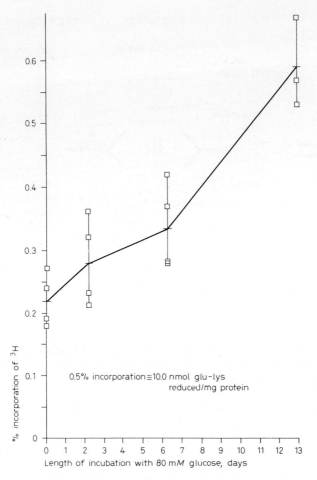

Fig. 2. Measurement of glycation in BSA by NaB³H₄ reduction.

Three types of BSA controls were examined over several months: (a) normal BSA batch-stored at −20 °C, (b) 13-day glycated BSA, and (c) 13-day buffer-treated BSA, but gave variable values for reducible group content (table I). The relationship between samples was inconsistent between experiments, and if the method is to be used it is comparative at best. Initial results in ketoacidotic patients suggested increased glycation and a reduction on treatment, but could not be consistently demonstrated, and must be regarded as unreliable. For 6

Table 1. Borohydride reduction of BSA

	Hydride incorporated, nmol/mg protein						
	Exp. 1	2	3	4	5	6	7
Untreated BSA	8.9	4.0	0.7	–	2.3	5.0	3.7
Glycated BSA[1]	7.0	9.2	8.1	3.9	4.5	12.0	11.4
Control BSA[1]	5.3	5.2	2.2	2.9	2.0	5.8	4.9

[1] BSA was incubated ±80 mmol/l glucose (glycated and control BSA respectively) at 37 °C for 13 days.

normals a comparative experiment gave a reducible group value of 3.11 ± 0.67 (mean ± SD) nmol H^-/mg protein compared to 3.9 ± 1.4 nmol/mg for 9 diabetics.

In summary, the borohydride-reducing method has been examined for estimating the glycation level in VLDL. Reproducibility was poor and increased levels were not shown in either stable or ketoacidotic diabetics.

References

1 Witztum, J.L.; Mahoney, E.M.; Branks, M.J.; Fisher, M.; Elam, R.; Steinberg, D.: Nonenzymatic glycosylation of low density lipoprotein alters its biologic activity. Diabetes 31: 283–291 (1982).

2 Curtiss, L.K.; Witztum, J.L.: A novel method for generating region-specific monoclonal antibodies to modified proteins. Application to the identification of human glucosylated low density lipoproteins. J. clin. Invest 72: 1427–1438 (1983).

3 Hui, D.Y.; Innerarity, T.L.; Mahley, R.W.: Lipoprotein binding to canine hepatic membranes. J. biol. Chem. 256: 5646–5655 (1981).

4 Rall, S.C.; Weisgraber, K.H.; Mahley, R.W.: Human apolipoprotein E. The complete amino acid sequence. J. biol. Chem. 257: 4171–4183 (1982).

5 Hunt, L.T.; Barker, W.C.; Dayhoff, M.O.: Miscellaneous proteins; in Dayhoff, Atlas of protein sequence and structure, vol. 5, suppl. 2, No. 17, pp. 257–268 (National Biomedical Research Foundation, 1976).

Dr. J. Reckless, Royal United Hospital, Bath BA1 3NG (England)

Mapping of Lipoprotein Particles with Monoclonal Antibodies in Diabetes and Atherosclerosis

C. Fievet[a], H. Parra[a], I. Luyeye[a], C. Demarquilly[a], P. Fievet[a], J.C. Fruchart[a], M. Bertrand[b], J.M. Lablanche[b], P. Drouin[c], P. Gross[c]

[a]Service de Recherche sur les Lipoprotéines et l'Athérosclérose (SERLIA) et Unité Inserm 279, Institut Pasteur, Lille; [b]Hôpital Cardiologique CHU, Lille; [c]Hôpital Jeanne d'Arc, Toul, France

Introduction

The most widely used classifications of plasma lipoproteins are based on nonspecific physicochemical properties as hydrated density, size or electrophoretic mobility. By the early sixties, the operationally defined polydisperse lipoprotein density classes or electrophoretic bands were accepted as the fundamental physicochemical and metabolic entities of the lipid transport system [1, 2]. This conceptual view was further enhanced and strengthened by clinical studies which related various dyslipoproteinemic states to particular density classes or electrophoretic patterns [3]. More recently, the discovery of several apolipoproteins and their wide distribution throughout the density spectrum has revealed a new dimension in the compositional complexity of plasma lipoproteins [4]. Apolipoproteins serve several important functions as structural components of lipoproteins, as activators or inhibitors of various enzymes which affect lipoprotein metabolism and as ligands for lipoprotein-receptor interactions.

Recent results from many laboratories have shown that major density classes or electrophoretic bands consist of several distinct lipoprotein particles rather than single homogeneous complexes [5]. The occurrence of apolipoproteins in nonequimolecular ratios provided further evidence that individual lipoprotein particles of the same density class could not have the same apolipoprotein composition. These

Table I. Criteria adopted for the selection of patients

Group 1	no coronary artery disease (grade 0 in all vessels) 42 male patients (mean age 48.7 ± 10.8 years)
Group 2	obstructive coronary artery disease (grade 2 or greater lesions in one or more vessels) 101 male patients (mean age 51.6 ± 7.8 years)
Group 3	reference population – healthy volunteers selected in a Center for Preventive Medicine 104 male patients (mean age 51.1 ± 8.2 years)

findings have led to the realization that apolipoproteins may be used as specific markers for the identification and the quantitation of lipoprotein particles.

According to this view, the purpose of our study was to map apolipoproteins B and A-I containing lipoprotein particles with monoclonal antibodies in diabetes and atherosclerosis.

Methods

Patient Selection

The subjects, 143 male patients referred for coronary angiography, were suspected for ischemic heart disease. Coronary angiography was performed according to the technique of *Judkins* [6]. Arteriograms were interpreted by two separate observers prior to any knowledge of the biochemical studies and the patients were divided into two groups on the basis of the angiographic results (table I). A third group of age-matched male controls was studied (table I).

Three other groups of patients were also selected: (1) 54 well-controlled diabetic male patients without atherosclerotic lesions; (2) 15 patients with distal arteriopathy without atherosclerotic lesions; (3) 6 patients with atherosclerotic lower-limb arteriopathy. The two first groups were compared with age- and sex-matched control subjects.

Analytic Methods

All biochemical studies were made on fresh sera stored at 4 °C for no longer than 48 h. Total cholesterol and triglycerides were determined by enzymatic methods [7, 8] adapted to centrifugal analysis. Apolipoproteins B and A-I levels were measured by noncompetitive enzyme-linked immunoassays according to figure 1 [9–11]. Polyclonal affinity-purified or monoclonal antibodies were adsorbed to the surface of microtiter plates. The solid-phase antibody was incubated with antigen, washed and then incubated

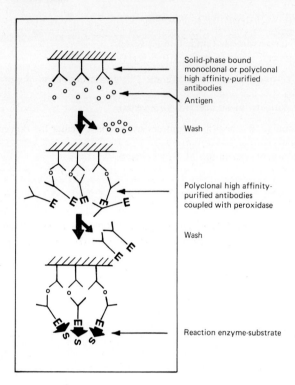

Fig. 1. Assessments of the apolipoproteins A-I and B by enzyme immunoassay using polyclonal or monoclonal antibodies.

with specific peroxidase-labelled affinity antibodies. After a last washing, the bound label was assayed, this providing a direct measurement of the antigen.

Polyclonal antisera were produced in goat or rabbits and affinity-purified antibodies were obtained as early reported [12]. Monoclonal antibodies were produced from mice by Research Center Clin Midy (Montpellier, France) and characterized [13]. They were designated as BL_3, BL_5 and BL_7 for those directed against apolipoprotein B and $2G_{11}$ for those directed against apolipoprotein A-I. Isolation of lipoprotein particles was done by monoclonal immunosorber [14].

The immunoaffinity chromatography offers a successful tool for the isolation of lipoprotein particles since the most suitable solid support was used. The most important criteria for this choice were the complete absence of nonspecific adsorption and suitable filtration properties. We used Sephadex G-25 (Pharmacia, Uppsala, Sweden) activated with CNBr as solid support of monoclonal antibody. After coupling, the immunosorber was incubated with fresh serum at room temperature for 2–4 h. The gel was then washed and retained fractions were eluted with mild elution agent (NaSCN 3 M) inducing no significant change in eluted lipoprotein particles. These were dialyzed, concentrated and tested for the presence of apolipoproteins by double immunodiffusion and enzyme-linked immunoassay using all antiapolipoproteins antisera disposable.

Results

Table II shows that lipoprotein profiles from well-controlled male diabetic patients without atherosclerotic lesions were not statistically different from those of age-matched normal subjects. The same type of results has been obtained when we have compared lipoprotein profiles of patients with distal arteriopathy without atherosclerotic lesions and those of age-matched normal subjects (table III). No significant differences can be observed.

In the lipoprotein profiles of 6 patients with atherosclerotic lower-limb arteriopathy, we observed (table IV) that in 5 patients (G.C., M.R., B.E., I.J., T.J.) the most characteristic abnormality was an accumulation of particles recognized by monoclonal BL_3. The ratio BL_3/B polyclonal which is usually 0.9 : 1 increased dramatically in these patients. The isolation of these particles was done as described in the 'Methods' section. According to immunoenzymatic analyses, the monoclonal BL_3 antibody recognizes particles containing apolipoproteins B, C-III and E. If we considered apolipoprotein A-I, we found a decrease in 2 patients but the values were the same with the monoclonal and polyclonal antibodies.

In a preliminary and parallel study on 143 angiographed patients, we have also observed that patients with coronary artery disease (table I, group 2) compared to patients with normal coronaries (table I, group 1) or age-matched controls (table I, group 3) had a significant increase in B particles specifically recognized by BL_3 monoclonal antibody (table V). The discriminatory capacity of these antibodies was found to be better than that of polyclonal antibodies but the other monoclonal antibodies (BL_5, BL_7) did not give a good discrimination.

Discussion

It has been reported recently that monoclonal antibodies are capable of recognizing specific epitope patterns and consequently separate

Table II. Lipoprotein profiles of well controlled male diabetic patients without atherosclerotic lesions (group 2): comparison with those of age-matched normal male subjects (group 1)

	Group 1 (n = 54)	Group 2 (n = 54)
Age, years	38.63 ± 13.88	38.72 ± 13.76
Cholesterol, mmol/l	5.61 ± 1.10	5.30 ± 1.32
Triglycerides, mmol/l	1.37 ± 0.74	1.24 ± 0.80
Apolipoprotein B, g/l		
Polyclonal	0.92 ± 0.32	0.77 ± 0.36
Monoclonal BL_3	1.04 ± 0.29	1.10 ± 0.49
Monoclonal BL_5	0.50 ± 0.18	0.35 ± 0.23
Monoclonal BL_7	1.00 ± 0.37	0.85 ± 0.44
Apolipoprotein A-I, g/l		
Polyclonal	1.74 ± 0.61	1.46 ± 0.66
Monoclonal $2G_{11}$	1.46 ± 0.47	1.24 ± 0.48

Results are expressed as mean ± SD.

Table III. Lipoprotein profiles of diabetic patients with distal arteriopathy without atherosclerotic lesions (group 2): comparison with those of sex- and age-matched normal subjects (group 1)

	Group 1 (n = 15)	Group 2 (n = 15)
Age, years	60.75 ± 7.43	63.75 ± 9.01
Cholesterol, mmol/l	6.06 ± 1.29	5.09 ± 1.08
Triglycerides, mmol/l	1.75 ± 1.20	1.51 ± 0.62
Apolipoprotein B, g/l		
Polyclonal	0.88 ± 0.31	0.76 ± 0.27
Monoclonal BL_3	1.26 ± 0.47	1.04 ± 0.49
Monoclonal BL_5	0.45 ± 0.35	0.40 ± 0.18
Monoclonal BL_7	0.96 ± 0.35	0.79 ± 0.28
Apolipoprotein A-I, g/l		
Polyclonal	1.54 ± 0.47	1.14 ± 0.32
Monoclonal $2G_{11}$	1.32 ± 0.33	1.02 ± 0.30

Results are expressed as mean ± SD.

Table IV. Lipoprotein profiles of 6 patients with atherosclerotic lower-limb arteriopathy

	G.C.	M.R.	B.E.	I.J.	T.J.	S.C.
Age, years	41	63	62	64	40	43
Cholesterol, mmol/l	5.20	7.05	8.95	8.08	11.06	8.93
Triglycerides, mmol/l	1.56	2.17	1.54	2.14	7.83	5.95
Apolipoprotein B, g/l						
Polyclonal	0.92	1.52	1.51	1.18	0.37	2.00
Monoclonal BL_3	1.83	1.98	2.43	1.69	2.55	1.72
Monoclonal BL_5	0.32	0.86	1.19	1.22	0.15	0.55
Monoclonal BL_7	1.26	1.48	3.02	2.23	0.62	2.20
Apolipoprotein A-I, g/l						
Polyclonal	1.40	1.29	1.31	0.74	1.10	1.23
Monoclonal $2G_{11}$	0.76	1.50	1.40	0.75	1.61	1.05
BL_3/B polyclonal	2.00	1.30	1.61	1.43	6.90	0.86

Table V. Lipoprotein profiles of male patients with coronary artery disease (group 2): comparison with those of male patients with normal coronaries (group 1) and age-matched male controls (group 3)

	Group 1 (n = 42)	Group 2 (n = 101)	Group 3 (n = 104)
Age, years	48.71 ± 10.84	51.62 ± 7.85	51.14 ± 8.22
Cholesterol, mmol/l	5.77 ± 1.37	6.51 ± 1.26	6.21 ± 1.13
Triglycerides, mmol/l	1.52 ± 0.72	1.82 ± 0.85	1.41 ± 0.63
Apolipoprotein B, g/l			
Polyclonal	0.93 ± 0.33	1.16 ± 0.42	0.99 ± 0.35
Monoclonal BL_3	1.10 ± 0.40	1.36 ± 0.42	1.06 ± 0.31
Monoclonal BL_5	0.63 ± 0.45	0.72 ± 0.46	0.52 ± 0.19
Monoclonal BL_7	1.01 ± 0.50	1.15 ± 0.57	0.90 ± 0.38
Apolipoprotein A-I, g/l			
Polyclonal	1.15 ± 0.31	1.16 ± 0.33	1.37 ± 0.38
Monoclonal $2G_{11}$	1.11 ± 0.36	1.13 ± 0.39	1.38 ± 0.55
BL_3/polyclonal	1.18 ± 1.21	1.17 ± 1.00	1.07 ± 0.89

Results are expressed ± SD.

particles that might be similar to conventional criteria. Using different monoclonal antibodies against apolipoprotein B coupled to nylon as immunosorbers, *Koren* et al. [14] have isolated five lipoprotein subspecies and proved that each corresponds to one lipoprotein particle.

Our results suggest that atherosclerotic patients have a significant increase in apolipoprotein B containing particles specifically recognized by BL_3 antibody. In order to confirm the close relationship between this parameter and atherogenesis, a prospective study linked to the WHO Monica Project is now starting in our laboratory.

Conclusion

Although determination of apolipoproteins in whole plasma provides new information and represents an absolute diagnostic criterion for some disorders of lipid transport, the immediate future goal should include the development of a relatively simple procedure for the quantification of lipoprotein particles. Monoclonal antibody techniques may be used in the future as a new approach to classify lipid transport disorders.

References

1 Gofman, J.W.; Delalla, O.; Glazier, F.; Nichols, A.V.: The serum lipoprotein transport system in health, metabolic disorders, atherosclerosis and coronary heart disease. Plasma *2:* 413–484 (1954).
2 Hatch, F.T.; Lees, R.S.: Practical methods for plasma lipoprotein analysis. Adv. Lipid Res. *6:* 1–68 (1968).
3 Fredrickson, D.S.; Levy, R.I.; Lees, R.S.: Fate transport in lipoproteins – an integrated approach to mechanisms and disorders. New Engl. J. Med. *276:* 32–77, 94–103, 148–156, 215–226, 273–281 (1967).
4 Alaupovic, P.; Lee, D.M.; McConathy, W.J.: Studies of the composition and structure of plasma lipoproteins. Distribution of lipoprotein families in major density classes of normal human plasma lipoproteins. Biochim. biophys. Acta *260:* 689–707 (1972).
5 Alaupovic, P.: The concepts, classification systems and nomenclatures of human plasma lipoproteins; in Lewis, Opplot, CRC handbook of electrophoresis. Lipoproteins: basic principles and concepts, pp. 27–46 (CRC Press, Boca Raton 1980).
6 Judkins, M.P.: Selective coronary arteriography. Radiology *89:* 815–819 (1967).
7 Fruchart, J.C.; Duthilleul, P.; Daunizeau, A.; Comyn, P.: Dosage du cholestérol total à l'aide d'une méthode enzymatique utilisant un monoréactif. Pharm. Biol. *24:* 227–229 (1980).

8   Ziegenhorn, J.; Bartl, K.; Deeg, R.: Improved kinetic method for automated determination of serum triglycerides. Clin. Chem. *26:* 973–979 (1980).
9   Fievet, C.; Koffigan, M.; Ouvry, D.; Marcovina, S.; Moschetto, Y.; Fruchart, J.C.: Noncompetitive enzyme-linked immunoassay for apolipoprotein B in serum. Clin. Chem. *30:* 98–100 (1984).
10 Fruchart, J.C.; Fievet, C.; Ouvry, D.; Koffigan, M.; Beucler, I.; Ayrault-Jarrier, M.; Du Barry, M.; Marcovina, S.: Enzyme-linked immunoassay on microtitre plates for human apolipoprotein B. Ric. Clin. Lab. *14:* 569–574 (1984).
11 Koren, E.; Puchois, P.; McConathy, W.J.; Fesmire, J.D.; Alaupovic, P.: Quantitative determination of human plasma apolipoprotein A-I by a noncompetitive enzyme-linked immunosorbent assay. Clinica chim. Acta (in press, 1985).
12 Axen, R.; Porath, J.; Ernback, S.: Chemical coupling of peptides and proteins to polysaccharides by means of cyanogen halides. Nature, Lond. *214:* 1302–1304 (1967).
13 Salmon, S.; Goldstein, S.; Beucler, I.; Ayrault-Jarrier, M.; Theron, L.: Discrimination entre LDL et VLDL par des anticorps monoclonaux anti-LDL. J. Gerli, Le Touquet, pp. 1–55 (1984).
14 Koren, E.; Solter, D.; Knight, C.; Alaupovic, P.: Isolation of apolipoprotein B containing lipoproteins by use of monoclonal antibodies. Circulation *68:* suppl. III, p. 217 (1983).

C. Fievet, MD, Service de Recherche sur les Lipoprotéines et l'Athérosclérose (SERLIA) et Unité Inserm 279, Institut Pasteur, 15, rue Camille Guérin, F-59019 Lille Cédex (France)

Effects of Fenofibrate on Receptor-Mediated and Receptor-Independent Low Density Lipoprotein Catabolism in Hypertriglyceridaemic Subjects[1]

C.J. Packard, M.J. Caslake, J. Shepherd

University Department of Pathological Biochemistry, Glasgow Royal Infirmary, Glasgow, UK

Treatment of hypertriglyceridaemic individuals with hypolipidaemic drugs causes an increase in the plasma LDL cholesterol concentration by mechanisms that are, as yet, unclear. The purpose of the present study was to examine the effect of fenofibrate, an effective triglyceride-lowering agent on LDL metabolism in severely hypertriglyceridaemic subjects.

Methods

Seven male subjects with plasma triglyceride levels in excess of 4.5 mmol/l and LDL cholesterol values of less than 3.5 mmol/l were investigated on two occasions, first while in the basal, untreated state and secondly during fenofibrate administration. At each time the composition and metabolism of LDL was determined according to a previously published protocol [1].

Results

Fenofibrate reduced plasma triglyceride levels in the 7 subjects by an average of 77% (table I). Plasma cholesterol also fell due to the reduction in very low density lipoprotein (VLDL). However, the level of the sterol rose in LDL and high density lipoprotein (HDL) by 40 and

---

[1] This work was supported by grants from the Scottish Hospital Endowments Research Trust (SHERT 673) and the Medical Research Council (G8111558 SA). Fenofibrate was the kind gift of Fournier Laboratories. *Joyce Pollock* provided excellent secretarial assistance.

Table I. Effects of fenofibrate on plasma lipids and lipoproteins

Subject	Triglyceride mmol/l		Cholesterol (mmol/l) in:				HDL_2 mg/dl		HDL_3 mg/dl	
			LDL		HDL					
	C	D	C	D	C	D	C	D	C	D
1	12.9	1.9	2.2	2.6	1.0	1.4	8	28	161	273
2	4.5	2.0	3.2	3.1	1.2	1.5	25	30	240	262
3	8.5	2.4	1.8	3.4	1.0	1.2	10	17	133	228
4	36.7	3.3	0.9	2.2	0.6	1.1	26	18	114	241
5	11.3	2.4	1.8	2.6	0.9	1.2	44	49	157	183
6	17.0	3.1	3.1	3.5	1.2	2.0	36	28	238	323
7	25.2	11.7	0.5	1.3	0.5	0.6	21	8	90	63
	16.6	3.8	1.9	2.7	0.9	1.3	24	25	162	225

C = control phase; D = fenofibrate treatment.

Table II. Effects of fenofibrate on plasma LDL metabolism

Subject	Apo-LDL mg/dl		Fractional clearance				Synthetic rate mg/kg/day	
			Receptor pools/day		Nonreceptor pools/day			
	C	D	C	D	C	D	C	D
1	71	74	0.13	0.22	0.35	0.26	13.6	14.3
2	82	82	0.09	0.17	0.31	0.25	13.9	13.7
3	46	92	0.27	0.18	0.33	0.16	11.1	12.5
4	22	66	0.15	0.15	0.90	0.27	9.4	11.1
5	80	87	0.10	0.17	0.20	0.18	9.6	12.2
6	70	81	0.26	0.21	0.46	0.21	20.1	13.6
7	25	58	0.36	0.21	0.84	0.32	12.6	12.4
	57	77	0.19	0.19	0.48	0.24	12.9	12.8

C = control phase; D = fenofibrate treatment.

43%, respectively, the latter being due to an increment in the HDL_3 subfraction.

During drug treatment, LDL became richer in cholesteryl ester (i.e. its mass contribution rose from 31 to 40%) and poorer in triglyceride (this fell from 6 to 2% of the mass). Hypolipidaemic therapy also affected the turnover of Apo-LDL in that while the synthetic rate of the apoprotein was virtually unaltered the mean FCR of LDL fell from 0.67 pools/day in the basal state to 0.43 pools/day on fenofibrate (table II). When the activities of the receptor and nonreceptor pathways were examined it was clear that the most consistent change occurred in the rate of LDL degradation by the receptor-independent route.

Discussion

Treatment of this group of subjects with fenofibrate caused their LDL to become enriched in esterified cholesterol (table I) so that its composition more closely resembles that of normal subjects [1]. *Deckelbaum* et al. [2] have suggested that in hypertriglyceridaemia, exchange of cholesteryl ester for triglyceride in the core of LDL allows the particle to be lipolysed to a greater degree than normal and so a smaller, denser lipoprotein is formed. Lowering VLDL triglyceride levels by treatment presumably limits these processes and so the LDL becomes normal in size and composition.

The reason for the occurrence of accelerated receptor-independent LDL catabolism in hypertriglyceridaemic subjects is not known, but therapy corrected this twofold increase so that the balance between receptor-mediated and nonreceptor catabolism was restored to normal. In fact the amount of LDL (in mg/kg/day) cleared by the receptors increased significantly in treated subjects. Fenofibrate treatment of these patients, therefore, markedly reduced plasma triglyceride, returned LDL structure, composition and metabolism towards normal, and raised HDL levels.

References

1 Packard, C.J.; McKinney, L.; Carr, K.; Shepherd, J.: Cholesterol feeding increases low density lipoprotein synthesis. J. clin. Invest. *72:* 45–51 (1983).
2 Deckelbaum, R.J.; Olivecrona, T.; Eisenberg, S.: Plasma lipoproteins in hyperlipidaemia: roles of neutral lipid exchange and lipase; in Carlson, Olsson, Treatment of hyperlipoproteinemia, pp. 85–93 (Raven Press, New York 1984).

C.J. Packard, PhD, University Department of Pathological Biochemistry, Glasgow Royal Infirmary, Glasgow G4 0SF (UK)

Pantethine versus Fenofibrate in the Treatment of Type II Hyperlipoproteinemia

Alfredo Postiglione, Paolo Rubba, Umberto Cicerano, Italia Chierchia, Mario Mancini

Institute of Internal Medicine and Metabolic Diseases, 2nd Medical School, University of Naples, Italy

Introduction

Long-term hypocholesterolemic treatment requires negligible side effects, but at the same time prolonged and substantial lipid-lowering effect with a decrease of plasma cholesterol value by at least 15% [1]. In order to avoid as much as possible unwanted clinical complications, interest has recently been focused on compounds derived from natural sources with lipid-lowering properties and theoretically free from clinical side effects, such as sulfated mucopolysaccharides, phospholipids and pantethine.

The aim of the present study was to compare the lipid-lowering effect of a natural compound as pantethine, the coenzyme A precursor, to that of fenofibrate, a well-known clofibrate derivative. The former compound has a well-known physiological role in lipid metabolism but has given variable results in clinical trials [2–9]. The latter has shown to be a potent cholesterol- and triglyceride-lowering agent with simultaneous decrease of low density lipoprotein (LDL)-cholesterol, very low density lipoprotein (VLDL)-triglyceride and Apo B [10–14].

Patients and Methods

Thirty-nine patients with asymptomatic primary type II hyperlipoproteinemia entered a 1-month open single-blind placebo period during which they received both fenofibrate and pantethine placebo capsules. Thirty of them, whose plasma cholesterol value was ≥ 280 mg/dl and triglyceride value ≥ 200 mg/dl at the end of this period, were randomly allocated into a fenofibrate-treated group (fenofibrate 300 mg/day and pantethine placebo) or into a panethine-treated group (pantethine 900 mg/day and fenofibrate placebo) in a controlled double-blind, double-dummy parallel study for a second month. Twenty-four (8 males, 16 males) (15 type IIA, 9 type IIB) ended the study. Plasma lipid, lipoprotein, apoprotein concentrations [15–17] and routine tolerance parameters were measured at the end of each study period.

Results

Fourteen patients received active fenofibrate treatment and 16 received pantethine. Six out of 30 patients, who were on one of the two active treatments, were dropouts. Two in the fenofibrate group interrupted the study because of minor allergic-type skin reactions and 1 because of gastric discomfort. One patient in the pantethine group was withdrawn for gastric discomfort and 2 refused the last blood sampling. Mean age, sex, type of hyperlipoproteinemia, mean body weight, cholesterol and triglyceride were comparable between both groups at entry into the treatment phase. No abnormal laboratory safety values were detected during any of the treatments.

Fenofibrate treatment decreased plasma cholesterol and triglyceride levels by 26% ($p < 0.001$) and 45% ($p < 0.01$) respectively, due to a reduction of LDL-cholesterol (-26%, $p < 0.001$) and of VLDL-triglyceride (-55%, $p < 0.05$). At the same time, 11% increase in high density lipoprotein (HDL)-cholesterol was observed with a parallel increase in plasma apoprotein A level ($p < 0.05$). Plasma values after pantethine were slightly affected after 1 month of treatment. Plasma cholesterol decreased by 4% (NS) in all patients and by 8.3% ($p < 0.01$) in type IIA only. Mean triglyceride, cholesterol content of LDL and HDL and Apo A were also unchanged. Plasma Apo B decreased not significantly by 7%. Table I shows the results at the end of 1 month of active treatment by the parametric test of analysis of covariance involving an adjustment for the placebo values. A marked and significant decrease of plasma cholesterol, triglyceride, Apo A was observed together with VLDL and LDL fractions.

Discussion

The present study shows that fenofibrate was more effective than pantethine in correcting the elevated lipids and lipoprotein abnor-

Table I. Comparative evaluation of the lipid-lowering effects (mg/dl) after 1 month of fenofibrate and pantethine (analysis of covariance). M ± SEM

	Fenofibrate	Pantethine	p
Cholesterol	258 ± 13	335 ± 12	0.001
Triglyceride	104 ± 16	172 ± 15	0.01
VLDL-triglyceride	63 ± 11	102 ± 10	0.05
LDL-cholesterol	193 ± 13	255 ± 12	0.005
HDL-cholesterol	50 ± 3	43 ± 3	NS
Apo A	221 ± 8	193 ± 7	0.05
Apo B	152 ± 9	176 ± 8	NS

malities of type II patients. The lipid-lowering effect obtained by pantethine in our study was poor, as often occurred in other trials with this drug, which showed very variable results [2–9]. Because various important investigations showed that the greater the reduction in plasma cholesterol, the greater was the benefit of coronary heart disease prognosis, the use of effective drugs should always be encouraged despite possible presence of unwanted side effects. The use of other drugs, although well tolerated and free from appreciable side effects, but with poor lipid-lowering activity, does not appear of any real advantage to the patient.

References

1 FDA – Bureau of Drugs: Endocrinologic and metabolic drugs advisory committee subcommittee. Regulatory criteria for evaluation of lipid-lowering agents (FDA, Rockville 1981).
2 Maggi, G.C.; Donati, C.; Criscuoli, G.: Pantethine: a physiological lipomodulating agent, in the treatment of hyperlipidemias. Curr. ther. Res. 3: 380–386 (1982).
3 Avogaro, P.; Bittolo Bon, G.; Fusello, M.: Effect of pantethine on lipids, lipoproteins in man. Curr. ther. Res. 3: 488–493 (1983).
4 Ranieri, G.; Chiarappa Balestrazzi, M.; De Cesaris, R.: Effect of pantethine on lipids and lipoproteins in man. Acta ther. 10: 219–227 (1984).
5 Maioli, M.; Pacifico, A.; Cherchi, G.M.: Effect of pantethine on subfractions of HDL in dyslipidemic patients. Curr. ther. Res. 2: 307–311 (1984).
6 Gaddi, A.; Descovich, G.C.; Noseda, G.; Fragiacomo, C.; Colombo, L.; Craveri, A.; Montanari, G.; Sirtori, C.R.: Controlled evaluation of pantethine a natural hypolipidemic compound in patients with different forms of hyperlipoproteinemia. Atherosclerosis 50: 73–83 (1984).

7   Dacol, P.G.; Cattin, L.; Fonda, M.; Mameli, M.G.; Feruglio, F.S.: Pantethine in the treatment of hypercholesterolemia: a randomized double-blind trial versus tiadenol. Curr. ther. Res. *2:* 314–322 (1984).
8   Goto, Y.; Hata, Y.; Kumagai, A.; Saito, Y.: A double-blind study on the effects of pantethine on hyperlipidemias. Abstr. VIIth Int. Symp. on Drugs Affecting Lipid Metabolism, Milan 1980.
9   Hata, Y.: Effect of pantethine on serum lipids and lipoproteins. A collaborative study on 871 patients with dyslipidemias. Abstr. VIIIth Int. Symp. on Drugs Affecting Lipid Metabolism, Philadelphia 1983.
10 Micheli, H.; Pometta, D.; Gustafson, A.: Treatment of hyperlipoproteinemia (HLP) type IIA with a new phenoxyisobutyric acid derivative, procetofen. Int. J. clin. Pharmacol. Biophys. Ther. Tox. *12:* 503–506 (1979).
11 Lehtonen, A.; Viikari, J.: Effects of procetofen on serum total cholesterol triglyceride and high density lipoprotein cholesterol concentrations in hyperlipoproteinemia. Int. J. clin. Pharmacol. Biophys. Ther. Tox. *12:* 534–538 (1981).
12 Viikari, J.; Solakivi-Jaakkola, T.; Lehtonen, A.: Effect of procetofen on apolipoprotein A_1 and B concentrations in hyperlipoproteinemia. Int. J. clin. Pharmacol. Biophys. Ther. Tox. *8:* 362–365 (1982).
13 Rubba, P.; Falanga, A.; Postiglione, A.; Patti, L.; Mancini, M.: Increase in lipoprotein lipase activity after procetofen (fenofibrate) treatment in primary hyperlipoproteinemia. Clin. Ther. Cardiovasc. *2:* 177–179 (1982).
14 Avogaro, P.; Bittolo Bon, G.; Belussi, F.; Pontoglio, E.; Cazzolato, G.: Variations in lipids and proteins of lipoproteins by fenofibrate in some hyperlipoproteinemia states. Atherosclerosis 47: 95–100 (1983).
15 Carlson, K.: Lipoprotein fractionation. J. clin. Path. *26:* suppl. 5, pp. 32–37 (1973).
16 Oriente, P.; Di Marino, L.; Mastranzo, P.; Iovine, C.; Patti, L.: Simultaneous determination of cholesterol and triglycerides in serum and in lipoprotein fractions with enzymatic automated methods. Clin. Eur. Symp., p. 387 (Piccin, Padova 1979).
17 Mancini, Y.; Carbonara, A.O.; Heremans, J.F.: Immunochemical quantification of antigens by single radial immunodiffusion. Immunochemistry *2:* 235–240 (1965).

Alfredo Postiglione, MD, Institute of Internal Medicine and Metabolic Diseases, 2nd Medical School, University of Naples, via S. Pansini 5, I-80131 Naples (Italy)

Influence of Carbohydrate Metabolism on Plasma Lipoprotein Levels

G. Riccardi, A. Rivellese, B. Capaldo, O. Vaccaro

Institute of Internal Medicine and Metabolic Diseases, Second Medical School, University of Naples, Italy

It is well known that major abnormalities of blood glucose metabolism, like severely decompensated insulin-dependent diabetes, have a strong effect on serum lipoprotein concentration [1]. Conversely, it is still under debate whether mild derangements of blood glucose regulation can influence blood lipid metabolism [2–4].

We have analyzed serum lipoproteins in individuals with impaired glucose tolerance (IGT) or mild noninsulin-dependent diabetes mellitus (NIDD) selected from the general population. Both IGT and NIDD individuals were contrasted with weight-matched and lean normoglycemic controls, selected from the same population. This procedure protects against possible selection bias and allows to differentiate the influence of hyperglycemia from the confounding effect of overweight on lipoprotein metabolism.

Impaired Glucose Tolerance

Sixty-nine individuals with IGT, selected among the employees of a telephone company, were compared with 130 lean and 125 weight-matched normoglycemic controls. Cases and controls were of both genders and had a similar age (range 40–59 years). Total serum cholesterol was very similar in IGT individuals and both control groups. Conversely, total serum triglyceride was substantially higher in the IGT group. Such a difference was, however, largely dependent on overweight, since no significant difference was found between IGT and

Table I. Plasma lipids in IGT individuals and controls (mean ± SD)

	n	Total cholesterol mg/dl	Total triglyceride mg/dl	HDL cholesterol mg/dl
IGT	65	214 ± 34	146 ± 59	39 ± 3
Weight-matched controls	125	219 ± 40	135 ± 64	40 ± 10
Lean controls	130	206 ± 40	117 ± 51*	–

* $p < 0.001$ vs. IGT.

weight-matched controls. HDL cholesterol (dextran sulfate-manganese chloride precipitation), not measured in lean controls, did not show any significant difference between the IGT and the weight-matched control group (table I).

Serum lipoproteins, separated by preparative ultracentrifugation, were very similar in IGT and weight-matched controls [5]. This confirms that the IGT condition is not associated, per se, to any lipoprotein abnormality. A slight triglyceride elevation might be found in IGT individuals as a consequence of overweight.

Noninsulin-Dependent Diabetes

Cholesterol and triglyceride concentration in total plasma and in different lipoproteins, separated by preparative ultracentrifugation, are shown in table II for NIDD patients and two groups (lean and weight-matched) of normoglycemic controls. They were all males and had a similar age (range 40–59 years).

Cholesterol in total plasma, VLDL and LDL was similar in NIDD patients and both control groups. Conversely, HDL cholesterol was substantially reduced in diabetic patients. This, however, was mainly due to the excess of overweight among diabetic patients, since the comparison with weight-matched controls did not show any significant difference.

Triglyceride in total plasma and in lipoproteins was higher in NIDD but, once again, this was largely due to the influence of body

Table II. Lipoprotein composition in male NIDD patients and controls (mean ± SD).

	Total plasma, mg/dl		VLDL, mg/dl		LDL, mg/dl		HDL, mg/dl	
	C	T	C	T	C	T	C	T
NIDD (n = 23)	200 ± 45	162 ± 92	29 ± 22	107 ± 88	130 ± 51	40 ± 15	42 ± 13	12 ± 7
Weight-matched controls (n = 23)	205 ± 49	139 ± 82	26 ± 20	102 ± 83	118 ± 35	21 ± 6**	47 ± 10	9 ± 4
Lean controls (n = 22)	196 ± 36	108 ± 50*	20 ± 13	74 ± 46	120 ± 34	18 ± 5*	53 ± 15*	9 ± 4

C = cholesterol; T = triglyceride.
* $p<0.01$; ** $p<0.001$ vs. NIDD.

weight with the exception of LDL fraction which was significantly increased in NIDD patients as compared with both control groups. This rules out the confounding effect of overweight, and suggests that an abnormal LDL composition is associated with NIDD.

The finding is also confirmed by a different type of analysis: LDL cholesterol plotted against LDL triglyceride concentration shows a firmly constant proportion of cholesterol and triglyceride in normoglycemic individuals, while NIDD patients have a disproportional elevation of triglyceride in this lipoprotein fraction (fig. 1).

Since triglyceride-enriched lipoproteins (intermediate density lipoproteins, IDL) are also separated in the LDL class, it is reasonable to assume that the increased concentration of LDL found in NIDD is accounted for by an increased IDL concentration. IDL is a catabolic product of VLDL whose final metabolic step takes place in the liver. At this site the increased blood glucose concentration, either directly or through apoprotein glycosylation, may interfere with triglyceride-rich lipoprotein metabolism [6].

Conclusions

The only lipoprotein abnormality which seems primarily associated with impaired blood glucose metabolism is a triglyceride enrichment in

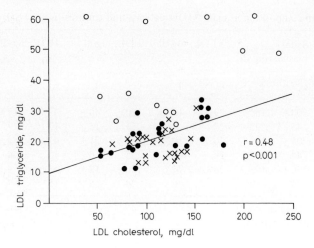

Fig. 1. Linear correlation between cholesterol and triglyceride in LDL. The correlation is significant only for lean (×) and weight-matched (●) normoglycemic controls but not for NIDD (○) patients.

LDL. This suggests an accumulation of IDL, an intermediate product of triglyceride metabolism.

References

1 Chance, G.W.; Albritt, E.C.; Adkins, S.M.: Serum lipids and lipoproteins in untreated diabetic children. Lancet *i:* 1126–1128 (1969).
2 Kannel, W.B.; Gordon, T.; Castelli, W.P.: Obesity, lipids and glucose intolerance. The Framingham Study. Am. J. clin. Nutr. *32:* 1228–1245 (1979).
3 Ballantyne, D.; Whithe, C.; Strums, E.A.; Lawrie, T.D.V.; Lorimer, A.R.; Monterson, W.G.; Morgan, H.G.: Lipoprotein concentrations in untreated adult-onset diabetes mellitus and the relationship of the fasting plasma triglyceride concentration to insulin secretion. Clinica chim. Acta *78:* 323–329 (1977).
4 Mancini, M.; Rivellese, A.; Rubba, P.; Riccardi, G.: Plasma lipoproteins in maturity-onset diabetes. Nutr. Metab. *24:* suppl. 1, pp. 65–73 (1980).
5 Capaldo, B.; Tutino, L.; Patti, L.; Vaccaro, O.; Rivellese, A.; Riccardi, G.: Lipoprotein composition in individuals with impaired glucose tolerance. Diabetes Care *6:* 575–578 (1983).
6 Weisweiler, P.; Jungst, D.; Schwandt, P.: Quantitation of apoliprotein E isoforms in diabetes mellitus. Hormone metabol. Res. *15:* 201–205 (1983).

Dr. G. Riccardi, Institute of Internal Medicine and Metabolic Diseases,
Nuovo Policlinico, Via S. Pansini 5, I-80131 Naples (Italy)

Apo B Degradation in Vascular Cells is Regulated by Metabolic Control of Diabetes

T. Koschinsky, C.E. Bünting, R. Rütter, F.A. Gries

Diabetes-Forschungsinstitut an der Universität Düsseldorf, FRG

Introduction

One feature common to both atherosclerosis and diabetes is an abnormal cholesterol and LDL metabolism. As there is no evidence for a LDL receptor defect or other inherited disorders of LDL metabolism in diabetes [1, 2], research has focused on acquired abnormalities of LDL catabolism in diabetes:

(1) Abnormal glycosylation of lysine residues of LDL-Apo B has been shown in type 1 diabetics with extremely severe hyperglycemia. The degradation of such glycosylated LDL-Apo B is decreased considerably in vivo and in vitro [3–6]. Tight control of diabetes normalized LDL degradation in vitro [6]. However, the relevance of the apoprotein glycosylation has been questioned. *Schleicher* et al. [7] found no effect on the catabolic rate of LDL from diabetic patients and *Kraemer* et al. [8] concluded that the LDL of patients with type 2 diabetes with moderate hyperglycemia are not modified sufficiently to alter their normal binding and degradation by human fibroblasts or to cause their uptake by mouse peritoneal macrophages.

(2) *Fielding* et al. [9] identified in vitro defects in cholesterol net transport from fibroblasts to plasma from poorly controlled type 2 diabetics, cholesterol esterification, and cholesteryl ester transfer to LDL. Effective control of hyperglycemia with insulin normalized these abnormalities. Their relevance for the cholesterol metabolism in diabetics remains to be clarified.

(3) Finally, *Wolinsky* et al. [10] reported that 4 weeks after induction of diabetes in rats lysosomal acid cholesterol esterase activity of

Table 1. Concentration range of pertinent blood constituents from poorly controlled, nonketotic, type 2 diabetics (NIDDM, DS−), from well-controlled type 2 diabetics (DS+) and from normal subjects (NS)

	NIDDM, DS− (n = 12)	NIDDM, DS+ (n = 5)	NS (n = 14)
Glucose, mg/dl	205–245	110–125	90–100
Cholesterol, mg/dl	180–225	126–209	142–159
Triglycerides, mg/dl	218–249	87–170	42–104
Insulin, µU/ml	22–34	18–28	7–14
Growth hormone, ng/ml	2.0–3.5	1.5–3.5	2.0–4.0
α_2-Macroglobulin, mg/dl	232–249	–	231–249
α_1-Antitrypsin, mg/dl	232–260	–	250–303

aortic smooth muscle cells decreased by 25% and returned to normal after insulin treatment.

As no information is available on the effect of metabolic control of type 2 diabetes on LDL apoprotein degradation in human vascular cells, we examined the effect of type 2 diabetes on lysosomal proteases. From the work of *Van der Westhuyzen* et al. [11] it is known that the lysosomal protease cathepsin D is the key enzyme that initiates Apo B degradation, followed by a synergistic hydrolysis with cathepsin B. Therefore, the effect of serum from poorly and from well-controlled type 2 diabetics has been studied on cathepsin D and B activity in human arterial smooth muscle cells and fibroblasts in vitro.

Methods

We used 12 pooled sera from overnight fasted noninsulin-dependent diabetic patients in poor metabolic control without ketosis (NIDDM, DS−), 5 pooled sera from noninsulin-dependent diabetics with optimal metabolic control (DS+) and 14 control sera pooled from healthy volunteers (NS). The concentration ranges of glucose, cholesterol, triglycerides, insulin, the protease inhibitors α_2-macroglobulin (α_2-m) and α_1-antitrypsin (α_1-a), and of growth hormone (GH) are given in table I.

Human skin fibroblasts and arterial smooth muscle cells were grown in Dulbecco's modification of Eagle's medium (DME) with the addition of 10% diabetic or control serum till confluency. Cells were washed free of serum, sonified and diluted to a constant protein concentration. Cathepsin D activity was measured at pH 3.6 using ^{125}I-LDL as substrate [12]. Activity was expressed as ng LDL protein degraded/µg DNA/h. Protease activity was also measured in the extracellular medium to control for secretion or loss

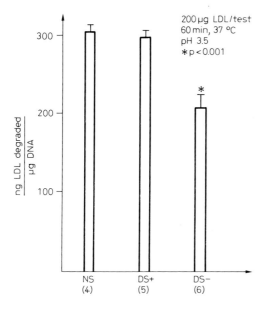

Fig. 1. Cathepsin D activity of human fibroblasts grown in DME + 10% normal serum (NS), serum from NIDDM with optimal (DS+) or poor metabolic control (DS−), using ^{125}I-LDL as substrate.

from the cells. The LDL were obtained from well-controlled, noninsulin-dependent diabetics. They were labeled with ^{125}I according to McFarlane. Cathepsin B activity was measured according to *Barret* [13].

Results and Discussion

The cathepsin D and B activity in fibroblasts was 3-fold higher than in arterial smooth muscle cells, but the effects of diabetic serum on LDL apoprotein catabolism were similar in both cell types. Diabetic serum from poorly controlled patients decreased LDL apoprotein degradation (20–200 µg/test) by cathepsin D by 16–54% when compared to normal serum ($p < 0.01$). Optimal metabolic control of diabetes abolished this effect (fig. 1). In contrast, diabetic serum showed no effect on cathepsin B. Its effect seems to be rather specifically directed to the key enzyme for Apo B degradation.

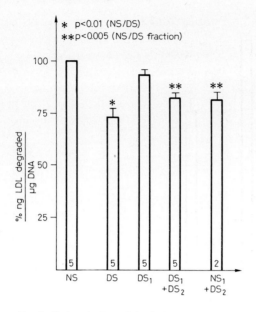

Fig. 2. Cathepsin D activity in human fibroblasts grown in DME + 10% normal serum (NS), serum from poorly controlled NIDDM (DS), dialyzed DS (DS$_1$), the recombined serum fractions DS$_1$ + the dialysate of DS (DS$_2$) or the recombined serum fractions from dialyzed normal serum (NS$_1$) + DS$_2$.

Preliminary studies on the effect of serum from poorly controlled diabetics on acid cholesterol esterase in cooperation with Dr. *Henze* from Munich showed a moderate decrease of acid cholesterol esterase activity by 5–24%. The decrease of cathepsin D activity was not due to loss or secretion of the enzyme into the extracellular medium. The effect on intracellular cathepsin D activity is related to low molecular weight components (< 12,000 daltons). A dialyzed diabetic serum fraction (> 12,000 daltons) did not differ anymore from normal serum while the recombination of the low molecular weight serum fraction with either the dialyzed diabetic or normal serum restored the original diabetes-specific effect (fig. 2).

Regarding the mechanisms, the following points should be considered: (1) A similar effect of serum from poorly controlled diabetics (DS−) on LDL apoprotein degradation occurs also in intact cultured fibroblasts [2]. (2) Since the molecular weight of the specific serum fraction is < 12,000 daltons, many small molecules like insulin or glucose have to be examined: insulin does not seem to be a primary candidate as the differences of the final concentration in the incubation medium were only 1–2 µU/ml, nor does glucose seem to be causative, as the differences of the final concentration in the medium were only 10–15 mg/dl. (3) Serum from poorly controlled type 2 diabetics contains increased amounts of a low molecular weight growth peptide [14]. As growth factors have been implicated in the regulation of intracellular protein catabolism [15] this diabetic growth factor could be involved in the inhibition of cathepsin D activity by diabetic serum.

References

1   Chait, A.; Biermann, E.L.; Albers, J.J.: Low density lipoprotein receptor activity in fibroblasts cultured from diabetic donors. Diabetes 28: 914–918 (1979).
2   Koschinsky, T.; Bünting, C.; Schwippert, B.; Gries, F.A.: Studies on inherited and acquired metabolic disorders in cultured arterial smooth muscle cells and fibroblasts. Int. J. Obes. 6: 91–104 (1982).
3   Witztum, J.L.; Mahoney, E.M.; Branks, M.M.; Fisher, M.; Elam, R.; Steinberg, D.: Nonenzymatic glucosylation of low-density lipoprotein alters its biologic activity. Diabetes 31: 283–291 (1982).
4   Kesaniemi, Y.A.; Witztum, J.L.; Steinbrecher, P.: Receptor-mediated catabolism of low density lipoprotein in man. Quantitation using glucosylated low density lipoprotein. J. clin. Invest. 71: 950–959 (1983).
5   Steinbrecher, P.; Witztum, J.L.; Kesaniemi, Y.A.; Elam, R.L.: Comparison of glucosylated low density lipoprotein with methylated or cyclohexanedione-treated low density lipoprotein in the measurement of receptor-independent low density lipoprotein catabolism. J. clin. Invest. 71: 960–964 (1983).
6   Lopes-Virella, M.F.; Sherer, G.K.; Lees, A.M.; Wohltmann, H.; Mayfield, R.; Sagel, J.; LeRoy, E.C.; Colwell, J.A.: Surface binding, internalization and degradation by cultured human fibroblasts of low density lipoproteins isolated from type 1 (insulin-dependent) diabetic patients: changes with metabolic control. Diabetologia 22: 430–436 (1982).
7   Schleicher, E.; Olgemöller, B.; Schön, J.; Dürst, T.: Non-enzymatic glycosylation of LDL does not alter its recognition by the high affinity receptor. Diabetologia 27: 328A (1984).
8   Kraemer, F.B.; Chen, Y.-D.I.; Cheung, R.M.C.; Reaven, G.M.: Are the binding and degradation of low density lipoprotein altered in type 2 (non-insulin-dependent) diabetes mellitus? Diabetologia 23: 28–33 (1982).
9   Fielding, C.J.; Reaven, G.M.; Fielding, P.E.: Human noninsulin-dependent diabetes: identification of a defect in plasma cholesterol transport normalized in vivo by insulin and in vitro by selective immunoadsorption of apolipoprotein E. Proc. natn. Acad. Sci. USA 79: 6365–6369 (1982).

10   Wolinsky, H.; Goldfischer, S.; Capron, L.; Capron, F.; Coltoff-Schiller, B.; Kasak, L.: Hydrolase activities in the rat aorta. I. Effects of diabetes mellitus and insulin treatment. Circulation Res. *42:* 821–831 (1978).
11   Van der Westhuyzen, D.R.; Gevers, W.; Coetzee, G.A.: Cathepsin-D-dependent initiation of the hydrolysis by lysosomal enzymes of apoprotein B from low-density lipoproteins. Eur. J. Biochem. *112:* 153–160 (1980).
12   Barret, A.J.: Cathepsin D and other carboxyl proteinases; in Barret, Proteinases in mammalian cells and tissues, p. 210 (Elsevier/North-Holland, Amsterdam 1977).
13   Barret, A.J.: Cathepsin B and other thiol proteinases; in Barret, Proteinases in mammalian cells and tissues, p. 181 (Elsevier/North-Holland, Amsterdam 1977).
14   Koschinsky, T.; Bünting, C.; Rütter, R.; Gries, F.A.: Sera from type 2 (non-insulin-dependent) diabetic and healthy subjects contain different amounts of a very low molecular weight growth peptide for vascular cells. Diabetologia *28:* 223–228 (1985).
15   Ballard, F.J.; Knowles, S.E.; Wong, S.S.C.; Bodner, J.B.; Wood, C.M.; Gunn, J.M.: Inhibition of protein breakdown in cultured cells is a consistent response to growth factors. FEBS Lett. *114:* 209–212 (1980).

Prof. Dr. F.A. Gries, Diabetes-Forschungsinstitut an der Universität Düsseldorf, Auf'm Hennekamp 65, D-4000 Düsseldorf 1 (FRG)

Diabetic Serum Growth Factor: A New Low Molecular Weight Growth Peptide for Arterial Smooth Muscle Cells of Platelet Origin

T. Koschinsky, C.E. Bünting, R. Rütter, R. Schütze, F.A. Gries

Diabetes Forschungsinstitut an der Universität Düsseldorf, FRG

Introduction

Abnormal cell growth is well established as an intrinsic part of the angiopathic process in both diabetic and nondiabetic atherosclerosis [1–4]. This abnormal growth may be under the influence of growth factors, present in sera of both diabetic and nondiabetic subjects. In type 2 diabetics, dialyzable serum growth factor(s) (MW < 12,000) stimulate increased growth and protein synthesis in human arterial smooth muscle cells and fibroblasts [5]. The increased growth-stimulating effect of this diabetic serum factor depends on the metabolic control as it occurs in poorly but not in well-controlled type 2 diabetics [6].

As the nature and origin of this diabetic serum growth factor (MW < 12,000) was unknown, its specificity for diabetes, its relative potency compared with various growth-stimulating hormones and its physicochemical properties have been examined. Platelets have been considered as a potential source as they release different growth peptides [7–11], and because in diabetes increased platelet aggregation has been related to poor metabolic control and the development of angiopathies [12, 13].

Materials and Methods

Subjects

Clinical data for the poorly controlled nonketotic type 1 and 2 diabetic patients and for the healthy subjects and the concentrations of pertinent blood constituents are shown in table I.

Table I. Clinical data and concentrations of pertinent blood constituents from poorly controlled, nonketotic, type 1 diabetics (IDDM), type 2 diabetics (NIDDM) and healthy subjects (controls) in the fasting state (mean ± SD)

	IDDM (n = 52)	NIDDM (n = 140)	Controls (n = 85)
Age, years	37 ± 16	62 ± 9	30 ± 6
Height, cm	169 ± 10	164 ± 8	172 ± 7
Weight, kg	67.9 ± 11.2	76.1 ± 14.8	64 ± 9
HbA$_1$, %	12.0 ± 1.5	11.9 ± 1.4	6.3 ± 0.6
Glucose, mg/dl	270 ± 52	250 ± 29	95.4 ± 0.7
Cholesterol, mg/dl	210 ± 49	228 ± 39	185 ± 23
Triglycerides, mg/dl	138 ± 86	240 ± 62	107 ± 36

Serum Fractionation

To remove low molecular weight compounds (MW < 3,500), serum was dialyzed for 48 h at 4 °C against 20 vol of NaCl (0.15 mol/l). The dialysate was discarded and the dialyzed serum fraction used as well as the undialyzed serum of the same batch for growth experiments. To examine the growth-stimulating effect of the low molecular weight fraction, serum was dialyzed (MW < 3,500) for 72 h at 4 °C against 10 vol of distilled water. The dialysate was mixed with Chelex 100 for 3 h, filtered, concentrated to dryness and reconstituted with distilled water to 8 ml. This fraction was applied to a Bio-Gel P-2 column. The resulting 9 eluate fractions were concentrated to dryness and reconstituted with the incubation medium.

Platelet Release Products

Washed platelets, derived in parallel and in amounts equivalent to the diabetic serum, were incubated with dialyzed platelet-poor plasma that was recalcified after the addition to the platelets (2 h, 37 °C). The platelet release products contained in the supernatant (20,200 g, 30 min, 4 °C) were further dialyzed and fractionated on the Bio-Gel P-2 column identically to serum to obtain the low molecular weight fraction.

Growth Assay

The growth of cultured human arterial smooth muscle cells and fibroblasts was examined as described previously [14].

Results and Discussion

Removal of the low molecular weight growth fraction (MW < 3,500) from sera of poorly controlled type 1 as well as type 2 diabetics reduced the growth-stimulating effect of the remaining serum growth factors on human fibroblasts by an average of 37% (p < 0.005) com-

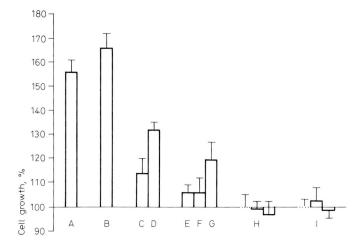

Fig. 1. Effect of 2 diabetic sera from poorly controlled type 2 diabetic patients (A) and their dialyzed sera (MW > 12,000) in combination with equivalent amounts of their dialysates (MW < 3,500) (B), human growth hormone (final medium concentration: 3 and 30 µg/l) (C, D), insulin (3, 30 and 300 mU/l) (E, F, G), glucagon (0.5, 5, and 50 µg/l) (H) or cortisol (8, 80 and 800 nmol/l) (I) on fibroblast growth. The cell number/plate on day 6, achieved with dialyzed diabetic serum, was taken as 100% (mean ± SD).

pared to diabetic serum. The recombination of the dialyzed diabetic serum (MW > 12,000) with its dialysate (MW < 3,500) restored the growth effect of the untreated serum. Removal of this growth-stimulating fraction from sera of healthy subjects also reduced the growth-stimulating effect on fibroblasts of the remaining serum growth factors, but only by an average of 8% ($p < 0.01$) compared with normal serum. This difference in effect on cell growth between diabetic and normal sera was due to differences in low molecular weight growth factor(s) as the dialyzed diabetic and normal sera no longer differed in their effect. Similar results were obtained with arterial smooth muscle cells.

The growth potency of the low molecular weight growth fraction (MW < 3,500) contained in 10% diabetic serum is 2–10 times higher than that of human growth hormone or insulin, added in amounts equivalent to 10% serum or physiological concentrations, whereas glucagon and cortisol did not show any growth activity above the basal level (fig. 1). This diabetic serum growth factor (MW < 3,500) is further characterized by: (1) linear dependence of growth stimulation over a

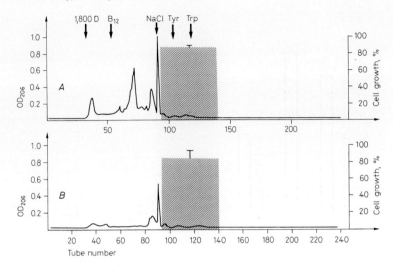

Fig. 2. Growth effect and elution profile from 3 dialysates of serum *(A)* and platelet release products *(B)* fractionated on a Bio-Gel P-2 column. Arterial smooth muscle cells were grown in parallel plates for 5 days till confluency in: DME + 5% dialyzed diabetic serum (MW > 12,000) + 9 column fractions (fraction 1: tube number 31–43; 2: 44–65; 3: 66–86; 4: 87–93 (salt fraction); 5: 94–139; 6: 140–162; 7: 163–187; 8: 188–213; 9: 214–240) from the dialysate (MW < 3,500) of serum *(A)* or platelet release products *(B)* of the same batch each derived from 25 ml blood/batch. The cell number/plate on day 5, achieved with dialyzed diabetic serum + the 9 recombined column fractions, is set 100% and compared to the growth effect of the single fractions 1–9 (mean + SD). The hatched area contains the growth-stimulating activity. Arrows mark different elution positions: vitamin B_{12} (MW 1,355), tyrosine (Tyr), tryptophan (Trp); the remaining 18 amino acids eluted between vitamin B_{12} and NaCl.

20 × concentration range, but only in combination with dialyzed serum (MW > 12,000); (2) reduction of the growth-stimulating activity to control levels by pretreatment with 2 different proteases: Serva pronase E (*Streptomyces griseus*; EC 3.4.24.4), Calbiochem protease (*Subtilisin carlsberg*; EC 3.4.4.16).

Diabetic platelets released during aggregation a dialyzable growth factor (MW < 3,500) that stimulated arterial smooth muscle cell growth to the same degree as the diabetic serum growth factor. Both growth factors eluted identically in only 1 Bio-Gel P-2 fraction behind the NaCl fraction indicating hydrophobic properties (fig. 2). These fractions comprised 84–88% of the total growth-stimulating capacity of the recombined fractions 1–9.

References

1 Ross, R.; Glomset, J.A.: Atherosclerosis and the arterial smooth muscle cell. Science *180:* 1332–1339 (1973).
2 Schwarz, S.M.; Gajdusek, C.M.: Growth factors and the vessel wall; in Spaet, Progress in hemostasis and thrombosis, vol. *6,* pp. 85–112 (Grune & Stratton, New York 1982).
3 Hauss, W.H.; Mey, J.; Schulte, H.: Effect of risk factors and antirheumatic drugs on the proliferation of aortic wall cells. Atherosclerosis *34:* 119–143 (1979).
4 Stout, R.W.: Diabetes and atherosclerosis – the role of insulin. Diabetologia *16:* 141–150 (1979).
5 Koschinsky, T.; Bünting, C.E.; Schwippert, B.; Gries, F.A.: Increased growth stimulation of fibroblasts from diabetics by diabetic serum factors of low molecular weight. Atherosclerosis *37:* 311–317 (1980).
6 Koschinsky, T.; Bünting, C.E.; Schwippert, B.; Gries, F.A.: Regulation of diabetic serum growth factors for human vascular cells by the metabolic control of diabetes mellitus. Atherosclerosis *39:* 313–319 (1981).
7 Ross, R.; Vogel, A.: The platelet-derived growth factor. Cell *14:* 203–210 (1978).
8 Oka, Y.; Orth, N.: Human plasma epidermal growth factor/β-urogastrone is associated with blood platelets. J. clin. Invest. *72:* 249–259 (1983).
9 Clemmons, D.R.; William, L.I.; Brown, M.T.: Dialyzable factor in human serum of platelet origin stimulates endothelial cell replication and growth. Proc. natn. Acad. Sci. USA *80:* 1641–1645 (1983).
10 King, G.L.; Buchwald, S.: Characterization and partial purification of an endothelial cell growth factor from human platelets. J. clin. Invest. *73:* 392–396 (1984).
11 Assoian, R.K.; Komoriya, A.; Meyers, C.A.; Miller, D.M.; Sporn, D.B.: Transforming growth factor-β in human platelets. J. biol. Chem. *258:* 7155–7160 (1983).
12 Born, M.N.: Platelet function in diabetes mellitus. Diabetes *27:* 342–350 (1978).
13 Colwell, J.A.; Lopes-Virella, M.; Halushka, P.V.: Pathogenesis of atherosclerosis in diabetes. Diabetes Care *4:* 121–133 (1981).
14 Koschinsky, T.; Bünting, C.E.; Rütter, R.; Gries, F.A.: Sera from type 2 (non-insulin-dependent) diabetic and healthy subjects contain different amounts of a very low molecular weight growth peptide for vascular cells. Diabetologia *28:* 223–228 (1985).
15 Rudermann, N.B.; Haudenschild, C.: Diabetes as an atherogenic factor. Prog. cardiovasc. Dis. *26:* 373–412 (1984).

Prof. Dr. F.A. Gries, Diabetes-Forschungsinstitut an der Universität Düsseldorf, Auf'm Hennekamp 65, D-4000 Düsseldorf 1 (FRG)

Tissue and Plasma Fibronectin in Diabetes

J. Labat-Robert, L. Robert

Laboratoire de Biochimie du Tissu Conjonctif, GR CNRS 40, Faculté de Médecine, Université Paris-Val-de-Marne, Créteil, France

Introduction

Diabetes of type I or II is accompanied by strong perturbations of the regulation of the biosynthesis of extracellular matrix macromolecules and especially of collagen, proteoglycans and structural glycoproteins [*Robert* et al., 1979]. Elastic fibers were also shown to be degraded in an accelerated fashion in major blood vessels and also in the skin [*Bouissou* et al., 1973]. In previous experiments we demonstrated in human conjunctival and skin biopsies an increased incorporation of ^{14}C-proline into polymeric collagen-bound hydroxyproline [*Kern* et al., 1976; *Miskulin* et al., 1979]. It was later shown that genetically diabetic mice of the KK strain also exhibit such dysregulations of extracellular matrix biosynthesis as exemplified by a significant increase of type III collagen biosynthesis in skin organ cultures without an increase of total collagen synthesis [*Kern* et al., 1979]. Very recently, similar results were obtained concerning the accelerated biosynthesis of type V collagen also in human diabetic skin biopsies [*Kern* et al., 1985]. In experiments undertaken in collaboration with Prof. *Leutenegger's* department at the Medical School in Reims, it could be shown that there is a considerable increase in fibronectin immunofluorescence in skin biopsies obtained from diabetic patients [*Leutenegger* et al., 1983; *Labat-Robert* et al., 1984a].

These results prompted us to explore in more detail the possible irregularities of fibronectin biosynthesis and metabolism both in tissues and in the plasma.

Table I. Plasma fibronectin in diabetic patients as compared to normals as a function of age and sex; average values ± SEM [Labat-Robert et al., 1984a]

		Age group					
		10–30 years		30–45 years		45–60 years	
		n	FN, mg%	n	FN, mg%	n	FN, mg%
Diabetics	M	34	19.0 ± 15	24	21.6 ± 11	14	25.4 ± 27
	F	24	19.0 ± 14	16	22.2 ± 16	5	19.0 ± 5
Controls	M	42	29.4 ± 11	14	33.2 ± 20	10	37.6 ± 31
	F	55	22.9 ± 16	26	28.0 ± 19	9	36.8 ± 29

Studies in Plasma Fibronectin

The plasma fibronectin levels were determined in diabetic patients, males and females, and the age and sex dependence of the plasma fibronectin values were compared with results obtained in the control population using a laser immunonephelometric procedure [Labat-Robert et al., 1981].

Table I shows the results which clearly show the difference in plasma fibronectin levels as a function of age and sex in diabetics as compared to controls. There is a considerable attenuation of the age-dependent increase of plasma fibronectin levels. In a normal population there is an exponential increase in plasma fibronectin with age [Labat-Robert et al., 1981]. This increase is nearly absent in the diabetic population. In the normal population, males have higher fibronectin values than females, but this difference tends to decrease with age. This sex-dependent difference was found to be completely absent in diabetics. These results indicate a deep-going perturbation in the regulation of plasma fibronectin synthesis and/or degradation.

Tissue Fibronectin in Diabetes

As mentioned in the Introduction, the first observation made during these studies was a conspicuous increase in the intensity of immune fluorescence observed in the skin biopsies obtained from

Table II. Incorporation of ^{35}S-methionine in normal (Swiss, C57 black), and diabetic (KK) mouse skin explants[1]

Strain	Animals	Medium	Tissue extract
Swiss	10	2.61 ± 0.12	2.26 ± 0.12
C57 black	10	2.69 ± 0.29	1.71 ± 0.16
KK	9	2.66 ± 0.13	3.57 ± 0.36[2]

[1] 24 h incubation, 37 °C, followed by immunoprecipitation of fibronectin from the SDS extracts and culture medium. Average values $\pm$ SEM of radioactivity incorporated in fibronectin as percentage of total incorporation.
[2] Compared to Swiss: $p < 0.01$; to C57 black: $p < 0.001$.

diabetic patients [*Leutenegger* et al., 1983; *Labat-Robert* et al., 1984a]. This increase concerns all the sites where fibronectin was present, but was the most significant in the dermoepidermal and capillary basement membranes as well as in the papillary dermis. These results suggested a local increase of tissue fibronectin. In order to control if local biosynthesis was really increased in diabetes, experiments were performed using genetically diabetic KK mouse skin explants. ^{35}S-methionine was incorporated in organ culture experiments in finely divided skin samples in Eagle's MEM during 24 h. After extraction with SDS 1% and dialysis, immune precipitation in the presence of a rabbit antiserum to human fibronectin and protein A, we could quantitate freshly synthesized radioactive fibronectin. Its identity was controlled by gel electrophoresis and fluorography.

Table II shows that in the diabetic KK mouse skin, there is a significant increase of incorporation of ^{35}S-methionine into freshly synthesized fibronectin as compared to control mice of the Swiss and C 57 black strains [*Labat-Robert* et al., 1984b]. No increase was observed in these experiments in the incorporation in fibronectin freely dissolved in the incubation medium, which can be interpreted as an efficient trapping of freshly synthesized fibronectin, by the extracellular matrix.

More recently, the above results could be confirmed and extended using fibroblast cell cultures derived from KK mouse skin and C 57 black or Swiss mouse skin [*Labat-Robert* et al., in preparation, 1985].

Discussion

The results obtained on human and mouse diabetic skin and fibroblasts strongly suggest a dysregulation of tissue and plasma fibronectin biosynthesis and/or degradation. As fibronectin was found to be increased at the level of the basement membranes, and as it is a PAS-positive glycoprotein, its increase can explain at least partially the strongly increased PAS positivity of the thickened capillary basement membranes in diabetic tissues [*Robert* et al., 1979]. The increased incorporation of ^{35}S-methionine in freshly synthesized fibronectin by KK mouse skin organ culture, by fibroblasts derived from skin, strongly supports this contention as well as our previous hypothesis claiming that the perturbation of extracellular matrix biosynthesis in diabetes is not restricted to the vascular wall but is present in most if not all the mesenchymal cells or at least in all cells synthesizing extracellular matrix macromolecules [*Robert* et al., 1979]. The decreased plasma fibronectin concentration found in human diabetics is more difficult to interpret; it is assumed that most if not all plasma fibronectin is synthesized in the liver [*Tamkun and Hynes*, 1983].

It is possible that this biosynthesis is slowed down in the liver, but it is equally possible that there is an increased trapping in the tissues of plasma fibronectin. Such trapping was shown to be possible by *Oh* et al. [1981]. These authors injected radioactive human plasma fibronectin in mice, and could demonstrate its penetration in the tissues. As we previously showed, a strong increase in type III collagen in diabetic tissues [*Kern* et al., 1979] and as it was proposed that type III collagen has the highest affinity for fibronectin [*Engvall* et al., 1978], it is plausible to assume that the increased type III collagen production represents an increased possibility of tissue fixation of plasma fibronectin in diabetic mesenchyme.

References

Bouissou, H.; Pieraggi, M.T.; Julian, M.; Douste-Blazy, L.: Cutaneous aging. Its relation with arteriosclerosis and atheroma; in Robert, Robert, Frontiers of matrix biology: aging of connective tissues – skin, vol. 1, pp. 190–211 (Karger, Basel 1973).

Engvall, E.; Ruoslahti, E.; Miller, E.J.: Affinity of fibronectin to collagens of different genetic types and to fibrinogen. J. exp. Med. *147:* 1584–1595 (1978).

Kern, P.; Moczar, M.; Robert, L.: Biosynthesis of skin collagen in normal and diabetic mice. Biochem. J. *182:* 337–345 (1979).

Kern, P.; Regnault, F.; Robert, L.: Biochemical and ultrastructural study of human diabetic conjunctiva. Biomedicine *24:* 32–38 (1976).

Kern, P.; Sebert, B.; Robert, L.: Increased type III/type I collagen ratios in diabetic human conjunctival biopsies. Clin. Physiol. Biochem. (in press, 1986).

Labat-Robert, J.; Leutenegger, M.; Llopis, G.; Ricard, Y.; Derouette, J.C.: Plasma and tissue fibronectin in diabetes. Clin. Physiol. Biochem. *2:* 39–48 (1984a).

Labat-Robert, J.; Potazman, J.P.; Derouette, J.C.; Robert, L.: Age-dependent increase of human plasma fibronectin. Cell Biol. int. Rep. *5:* 696–973 (1981).

Labat-Robert, J.; Potazman, J.P.; Robert, L.: Modification of the age-dependent increase of plasma fibronectin in cancer patients. Biochem. Soc. Trans. *12:* 660 (1984b).

Leutenegger, M.; Birembaut, P.; Poynard, J.P.; Eschard, J.P.; Ricard, Y.; Caron, Y.; Szendroi, M.; Labat-Robert, J.: Distribution of fibronectin in diabetic skin. Pathol. Biol. *31:* 45–48 (1983).

Miskulin, M.; Tixier, J.M.; Robert, L.: Rôle du tissu conjonctif dans le diabète. Méthode d'étude et résultats chez les diabétiques avant et après traitement. Rev. fr. Endocr. clin. *20:* 21–27 (1979).

Oh, E.; Pierchbacher, M.; Ruoslahti, E.: Deposition of plasma fibronectin in tissues. Proc. natn. Acad. Sci. USA *78:* 3218–3221 (1981).

Robert, A.M.; Miskulin, M.; Tixier, J.M.; Rivault, A.; Robert, L.: Biopsy technique for the exploration of the relative rates of biosynthesis of matrix macromolecules in normal and diabetic connective tissues; in Robert, Boniface, Robert, Frontiers of matrix biology, vol. 7, pp. 314–323 (Karger, Basel 1979).

Tamkun, J.W.; Hynes, R.O.: Plasma fibronectin is synthesized and secreted by hepatocytes. J. biol. Chem. *258:* 4641–4647 (1983).

J. Labat-Robert, PhD, Laboratoire de Biochimie du Tissu Conjonctif, GR CNRS 40, Faculté de Médecine, Université Paris-Val-de-Marne, 8 rue du Général-Sarrail, F-94010 Créteil Cédex (France)

Some Cellular and Molecular Aspects of the Vascular Complications of Diabetes

M. Miskulin, L. Robert, A.M. Robert

Laboratoire de Biochimie du Tissu Conjonctif, GE CNRS 40,
Université Paris XII, Faculté de Médecine, Créteil, France

It is well known that in diabetic patients arteriosclerosis develops much faster and is more severe. This speeding up of the arteriosclerotic process is often overshadowed by the development of microangiopathy with its severe complications. The involvement of the whole vascular system in the complications of diabetes is accompanied by the accelerated loss of elastic lamellae from arteries and the thickening of the basement laminae of the capillaries.

In a previous study on skin biopsies of patients with diabetic microangiopathy, we could establish that the biosynthesis of collagen is impaired and that long-term flavonoid treatment could improve clinically microangiopathy and normalize also collagen biosynthesis [1]. The purpose of the present investigation was to study some of the cellular and molecular perturbations involved in the vascular complications of diabetes and also to see if insulin and flavonoid drugs which are effective in patients, can also act at the cellular level. We chose as a model the streptozotocin-induced diabetes in rats. We shall report here briefly on results obtained on aorta smooth muscle cells (SMC). The details of the method used are described elsewhere [in preparation].

Results

Cell Attachment

When freshly trypsinized normal SMC are plated in a serum-free medium, 80–90% of the cells attach to the culture flask within 40 h.

Table I.

Cells and drug concentration	Radioactivity in total proteins cpm/µg DNA · 10^{-3}	Radioactivity in collagen (OH-Pro) cpm/µg DNA · 10^{-3}
Normal	21.2± 8.1	7.8±4.5
Normal + A 2 µg/ml	15.8± 9.3	3.9±1.8
Normal + A 20 µg/ml	6.8± 0.9	1.3±5.2
Diabetic	58.9±27.2	15.3±6.9
Diabetic + A 2 µg/ml	42.3±18.3	10.2±3.3
Diabetic + A 20 µg/ml	12.3± 1.58	5.2±5.3

Estimation of total protein and collagen synthesis by ^{14}C-proline incorporation. Effect of Anthocyanosides (A). The given results are the average of 8 independent experiments ± SEM.

Diabetic SMC attain the same percentage of attachment only after more than 60 h. When the serum-free medium is supplemented with 20 µg/ml of anthocyanosides of *Vaccinium myrtillus* (A), the attachment of the cells, normal or diabetic, is completely inhibited. When the same drug was used in a lower concentration, 2 µg/ml, only 55% of the normal and 35% of the diabetic cells attached to the flasks. This suggests that the drug may react with cell membrane components involved in attachment. Once the cells attached to the culture flasks, they cannot be detached by simple washing. A proteolytic enzyme must be used (trypsin or collagenase). 10–20 µg/ml of collagenase added to the flasks during 5 min could detach 50% of the normal SMC. When A is added to the flasks containing already attached cells, for 2 µg/ml of A 40–60 µg/ml, for 20 µg/ml of A 100–150 µg/ml of collagenase was necessary to detach 50% of the normal SMC. These results also indicate an interaction between A and the cell membrane: the drug may reinforce the cell attachment by modifying one or several of the cell surface components which participate in the attachment.

Table II.

Cells and drugs	Elastolytic activity	
	intracellular	extracellular
Normal	36 ± 13	0.5 ± 16
Normal + insulin	16 ± 6	0.32 ± 0.12
Normal + A	28 ± 8	0.34 ± 0.10
Diabetic	206 ± 68	3.74 ± 1.3
Diabetic + insulin	40 ± 12	0.40 ± 0.1
Diabetic + A	130 ± 46	1.14 ± 0.48

Estimation of the elastolytic activity of the SMC. Results are given in μM of N-Suc(Ala)$_3$-pNA hydrolyzed/min/2.10^6 of SMC. Concentration of insulin: 100 μM/ml; concentration of Anthocyanosides (A): 2 µg/ml. Indicated results are the average of 6 parallel determinations $\pm$ SEM.

Biosynthesis of Macromolecules

The biosynthetic activity measured by ^{14}C-proline incorporation is increased in diabetic cells for proteins in general and for polymeric collagen in particular. Protein synthesis was found to be linear during 24 h. Diabetic cells synthetize during this time twice as much collagen than normal cells and more than the double of protein. When A was added to the medium, the incorporation of labelled proline decreased in a dose-dependent way both in proteins and in collagen (table I). This finding confirms at the cellular level the results we found in our clinical studies [1]. Insulin does not produce the same effect: added to the medium in a concentration of 100 µU/ml, it has no significant effect on collagen synthesis in normal SMC, and in diabetic cells the reduction is of the order of 20% against 60% for the A.

Pinocytotic Activity

We also studied the pinocytotic activity of the SMC by uptake of radiolabelled sucrose. We found the uptake of this sugar was nearly doubled in diabetic SMC compared to the normal controls. This difference seems to indicate an alteration of the cell membrane in diabetic SMC. When A is added to the medium, the uptake of sucrose decreases in normal as well as in diabetic cells, probably as a consequence of the interaction of the drug with the cell surface.

Proteolytic Activity

Finally, we studied the proteolytic activity of the aorta SMC with collagen and elastin as substrates. Both of these fibrous proteins were much more degraded by the diabetic cells, which synthetized more proteases than normal SMC, and exported part of them in the medium. This was not the case with normal cells (table II). When insulin is added to the medium, the increase of the proteolytic activity of diabetic cells is suppressed, while A produces only a slight reduction.

Discussion and Conclusions

The normalization of the proteolytic activity in diabetic cells by insulin, together with the possible role of the elastase-type protease of the SMC of the aorta in the development of atherosclerosis, suggests that an insulin-dependent mechanism may well be involved in the accelerated loss of elastic lamellae in the arterioatherosclerotic disease in diabetes. The pharmacological effect of flavonoid drugs which act on the vascular wall consists in the correction of pathologically increased permeability and vascular fragility, judged by the reduction of the extravascular leakage of fluorescent or radiolabelled permeability tracers and petechiometry. Such effects of A were observed experimentally [2] as well as in clinical studies [3]. In our previous studies [1] we demonstrated the possibility of the correction of clinical symptoms of microangiopathy and the altered biosynthesis of collagen.

The results of the present study are coherent with the clinical findings on several points: the reduction of the pinocytotic activity of SMC, together with the crosslink-increasing effect on collagen [2], can represent a factor of the permeability-reducing effect of A.

The increased biosynthesis of proteins and particularly that of collagen and fibronectin by diabetic SMC and fibroblasts [4] is coherent with our findings on skin biopsies of diabetic patients. The reduction by A of this increased biosynthetic activity, observed in human diabetics, was also observed in SMC cultures. Both effects of A, on permeability and on collagen biosynthesis, may well be related to an interaction of A with extracellular matrix macromolecules on the cell surface.

References

1 Lagrue, G.; Robert, A.M.; Miskulin, M.; Robert, L.; Pinaudeau, Y.; Hirbec, G.; Kamalodine, T.: Pathology of the microcirculation in diabetes and alterations of the biosynthesis of intercellular matrix macromolecules; in Robert, Boniface, Robert, Frontiers of matrix biology, vol. 7, pp. 324–335 (Karger, Basel 1979).
2 Robert, A.M.; Godeau, G.; Moati, F.; Miskulin, M.: Action of anthocyanosides of *Vaccinium myrtillus* on the permeability of the blood brain barrier. J. Med. *8:* 321–332 (1977).
3 Lagrue, G.; Behar, A.; Baillet, J.; Zhapova, F.; Gumpelson, A.: Effet d'un flavonoide sur la perméabilité capillaire au cours des œdèmes idiopathiques orthostatiques. Gaz. méd. Fr. *86:* 1964–1967 (1979).
4 Labat-Robert, J.; Leutenegger, M.; Llopis, G.; Ricard, Y.; Derouette, J.C.: Plasma and tissue fibronectin in diabetes. Clin. Physiol. Biochem. *2:* 39–48 (1984).

A.M. Robert, MD, Laboratoire de Biochimie du Tissu Conjonctif, Faculté de Médecine, Université Paris-Val-de-Marne, 8, rue du Général-Sarrail, F-94010 Créteil Cédex (France)

Pathogenesis and Clinical Relevance of Mönckeberg's Medial Calcinosis

N. Zöllner, H.S. Füessl, F.D. Goebel

Medizinische Poliklinik der Universität München, FRG

Aetiology and pathogenesis of medial calcinosis are unknown. Earlier investigators reported on an increased incidence in diabetes mellitus, osteoporosis, uraemia, hyperparathyroidism and intoxication with vitamin D. Only in 1982, *Edmonds* et al. demonstrated a significant correlation between medial calcinosis and diabetic polyneuropathy, independent of age and duration of diabetes. We were interested in the question whether in diabetes medial calcinosis is a consequence of hyperglycaemia or of diabetic neuropathy. We therefore studied patients with lumbar sympathectomy with or without diabetes.

In order to demonstrate medial calcinosis, X-rays with soft tissue technique of the anterior part of both feet were taken. The roentgenographs were evaluated by a radiologist without knowledge of the clinical findings.

The patient material consisted of 60 patients in whom lumbar sympathectomy was performed in the years 1974–1977 because of restricted arterial perfusion of the lower extremity. In 27 cases sympathectomy was bilateral, in 33 cases sympathectomy was unilateral. 60 age- and sex-matched patients served as controls. In 55 (91,7%) of 60 sympathectomized patients a one- or both-sided medial calcinosis could be found. The correlation is highly significant ($p < 0.01$). 93% of the patients operated on both sides had a medial calcinosis on both sides. Of 33 patients operated only unilaterally, 29 (88%) showed a medial calcinosis on the operated side. In 23 (80%) of these 29 patients the calcinosis was limited to the side of the operation. Of the controls, 7 (11.2%) showed the characteristic signs of medial calcinosis.

We could not find that the correlation between diabetes and medial calcinosis is as clear as the correlation between medial calcinosis and sympathectomy. In our patient group, 19 patients had diabetes and 17

Table I. Incidence of medial calcinosis in sympathectomized patients (unilateral or bilateral) and in age- and sex-matched controls

	Sympathectomy		Controls
	bilateral	unilateral	
Number of cases	27	33	60
Bilateral medial calcinosis	25	6	5
Unilateral medial calcinosis	1	23[1]	2
No medial calcinosis demonstrable	1	4	53

[1] Only on the side of operation.

of these had medial calcinosis. On the other hand, 38 of 41 patients without diabetes also had medial changes. We found that the length of the vascular calcification was longer in diabetic patients than in patients with normal metabolism. However, we were unable to show statistical significance for this difference. The average age of our patients was 67,5 years. The ratio between men and women was 7.5:1.

As shown by *Edmonds* et al. [1982] in young diabetic patients, there is a highly significant association between diabetic polyneuropathy and medial calcinosis. However, there was very little correlation between this arterial disease and microangiopathy. None of our patients had manifest signs of retinopathy or nephropathy. Therefore we think that Mönckeberg's medial calcinosis is associated with a disturbance of the autonomous innervation of the arteries. It is not a metabolic disease per se which produces the medial calcinosis but the severe neural impairment.

We do not want to finish without mentioning that Mönckeberg's calcinosis has important clinical aspects. In the presence of Mönckeberg's sclerosis, arterial perfusion cannot be measured according to the principle of Doppler. This is quite important for correct blood pressure measurements in the upper extremity. Fortunately, Mönckeberg's calcinosis in the arm is quite rare.

Reference

Edmonds, M.E.; Morrison, N.; Laws, J.W.; Latkins, P.J.: Medial arterial calcification and diabetic neuropathy. Br. med. J. *284:* 928–930 (1982).

Prof. N. Zöllner, Medizinische Poliklinik der Universität München,
D-8000 München (FRG)

Subject Index

Acyl coenzyme A:cholesterol transferase (ACAT) 64
Adipose tissue lipoprotein lipase activity 124
Antibodies against glycosylated LDL 69
Apoprotein(s) (apolipoproteins, apo)
  chromosome localisation 27
  DNA sequences 26
  in diabetes mellitus 14
  lipoprotein composition 13
  of human plasma 13
  signal peptides 26
Apoprotein A-1
  in diabetes mellitus 17
  genetic mutants 27, 28
  quantitation by ELISA 136
Apoprotein A-II in diabetes mellitus 17
Apoprotein B
  carbohydrate composition 53
  in diabetes mellitus 17
  quantitation by ELISA 136
Aproprotein C in diabetes mellitus 22
Apoprotein C-III carbohydrate composition 55
Apoprotein E
  and type III hyperlipidaemia 29
  carbohydrate composition 56
  genetic variations 27, 28
  in diabetes mellitus 20
  influence on plasma lipid levels 30
  isoforms 29
  sialylated isoforms in diabetes 112

Apoprotein E-I 56
Atherosclerosis index 17, 106, 107

Bezafibrate
  effect
    on IDL metabolism 90
    on LDL metabolism 90
    on VLDL metabolism 90
  hypocholesterolaemic effect 89
  hypolipidaemic effect in diabetes mellitus 21
Biguanides 50
Borohydride reduction to estimate VLDL glycosylation 132

Cathepsin B and D activity
  effect of diabetic sera 154, 155
  in arterial smooth muscle cells 154
  in fibroblasts 154
Cholestyramine 76
  effect
    on LDL cholesterol 77, 80
    on serum cholesterol 77, 80
  treatment
    and cardiovascular disease 78, 79
    and gastrointestinal side effects 79
Chylomicronaemia 49
  and diet 50
Chylomicron remnants and atherogenesis 29

Subject Index

Clofibrate
  derivatives 86
  effect
    on HDL cholesterol 89
    on LDL cholesterol 85
    on VLDL levels 85, 86
  mechanism of action 87, 88
C-peptide and VLDL triglyceride synthesis 49

Diabetes, insulin-dependent (IDDM or type 1)
  apoprotein levels 17, 105, 106, 116
  familial hypertriglyceridaemia 74
  insulin gene polymorphism 39
  Lp(a) levels 19
Diabetes, insulin-independent (NIDDM or type 2)
  apoprotein levels 18, 105, 106
  insulin gene polymorphism 38
  lipoprotein lipid levels 16, 105, 150
  Lp(a) levels 19
Diabetes mellitus
  cardiovascular disease 3
  cerebrovascular disease 3
  insulin gene polymorphism 38
  levels
    of HDL cholesterol 4, 16
    of LDL cholesterol 4, 17
    of triglycerides 4
    of VLDL triglycerides 16
Diabetic serum growth factor, effect
  on arterial smooth muscle cells 161
  on fibroblasts 161
Diet, effect
  on LDL cholesterol 76
  on plasma cholesterol 76
  on VLDL synthesis 50

Enzyme-linked immunoassay (ELISA), for quantifying apoproteins 136
Epidemiological studies
  cholesterol lowering and coronary heart disease 74–78
  glucose intolerance and vascular disease 1–11

  hyperglycaemia and coronary heart disease 92–97

Familial hypercholesterolaemia
  heterozygous 65, 146
  homozygous 65
Fenofibrate, effect
  on apoprotein A-1 146
  on HDL cholesterol 143, 146
  on LDL metabolism 144
  on plasma cholesterol 143, 146
  on plasma triglycerides 143, 146
Fibronectin
  in diabetic plasma 165
  in diabetic skin tissue 166
Framingham study 1
Free fatty acids (FFA)
  and VLDL secretion 49
  effect of clofibrate treatment 86

Glibenclamide and lipid levels in NIDDM 116
Glucose intolerance
  and cardiovascular disease 3, 93
  and cerebrovascular disease 3
  and HDL cholesterol 150
  and serum lipid levels 149
Glucose loading test 93
Glucose plasma levels and cardiovascular disease 4
Glycaemia
  and triglyceride levels 46
  and VLDL levels 46
Glycosurea
  and cardiovascular disease 3
  and cerebrovascular disease 3
Glycosylation
  and lipoprotein secretion 55
  enzymatic 53
    of apoprotein B 53
  non-enzymatic 53, 131
    of apoprotein B 22
    of HDL 63, 69
    of LDL 63, 67

Subject Index

of lipoproteins 56
of VLDL 131
Gout
  and apoprotein gene polymorphism 122
  and hypertriglyceridaemia 121

Haemoglobin
  A1 117
  A1c 56
Hepatic lipase and fibrate treatment 88
High density lipoproteins (HDL)
  cholesterol
    in IDDM 107, 116
    in NIDDM 107, 116, 150
  effect of glycosylation on metabolism 58, 70
  half-life in plasma 57, 70
  in diabetes mellitus 16
3-Hydroxy-3-methylglutaryl-conezyme A (HMG CoA) reductase, effect of bezafibrate treatment on activity 89
Hyperglycaemia and coronary disease 92
Hypertriglyceridaemia and fenofibrate treatment 142

IDDM, see Diabetes, insulin-dependent
Immunoaffinity chromatography, to purify lipoprotein particles 137
Insulin
  and coronary heart disease 95
  and LDL catabolism 67
  gene polymorphism 35, 36
  gene structure 34
  genotype
    and hypertriglyceridaemia 39
    and insulin secretion 42
Intermediate density lipoproteins (IDL), in NIDDM 151
Intralipid clearance
  in fibrate treatment 88
  in IDDM 47
Intravenous fat tolerance test (IVFTT)
  activity of lipoprotein lipase 125
  and lipoprotein lipid levels 125, 126
  in NIDDM 125

Lecithin:cholesterol acyltransferase (LCAT), in diabetes mellitus 19
Lipoproteins
  apoprotein composition 13
  glycosylation and atherosclerosis 59, 63
Lipoprotein (a) [Lp(a)]
  carbohydrate composition 56
  in diabetes mellitus 17
Lipoprotein lipase
  activity in IDDM 47
  during fibrate treatment 88
  effect of insulin treatment 47
  activity in NIDDM
    adipose tissue 49, 125
    effect of insulin treatment 50
    postheparin plasma 49
    skeletal muscle 125
Low density lipoproteins
  carbohydrate composition 54
  carbohydrate content in diabetics 58
  cholesterol
    in IDDM 106, 117
    in NIDDM 106, 117
  composition in NIDDM 151
  desialylation and catabolism 54
  effect of glycosylation on metabolism 58, 64, 65
  glycosylation
    and fractional catabolic rate 66
    and receptor binding 59, 64, 65
  half-life in plasma 58
  methylated 65
  receptor-dependent
    pathway 64, 65, 68, 144
  receptor-idependent pathway 64, 65, 68, 144

Macrophage, uptake
  of glycosylated HDL 70
  of glycosylated LDL 65
Metabolic control of diabetes
  and plasma apoprotein levels 16
  and plasma VLDL levels 46, 47
Metabolism
  of apoprotein B 85–91
  of HDL 58, 69

Subject Index

Metabolism (continued)
  of LDL 58, 66, 142
  of VLDL triglycerides
    in IDDM 45
    in NIDDM 47
Mönckeberg's medial calcinosis 173
Monoclonal antibodies
  against glucitollysine 69
  and apoprotein epitope expression
    in coronary artery
      disease 139
    in diabetes 138
    in lower limb atherosclerosis 138
  assay
    of apoprotein A–I
    of apoprotein B

NIDDM, see Diabetes, insulin-independent

Pantethine, effect
  on apoprotein A–1 plasma levels 146
  on apoprotein B plasma levels 146
  on HDL cholesterol 146
  on LDL cholesterol 146
  on plasma cholesterol levels 146

Receptor
  for apoprotein E in liver 29
  LDL gene localisation 27

Schiff base 57, 62, 131
Sialic acid and lipoprotein metabolism 54
Skeletal muscle lipoprotein lipase activity 124

Smooth muscle cell cultures
  from diabetic rats 169
  from normal rats 169
Sulphonyl ureas 50
  and lipoprotein apoprotein levels 107
  and lipoprotein lipid levels 107

Tissue culture of arterial smooth muscle cells 155, 159, 169
Type 1 diabetes, see Diabetes, insulin-dependent
Type 2 diabetes, see Diabetes, insulin-independent

Very low density lipoproteins (VLDL)
  and ketoacidosis 46
  correlation
    with HbA1c 46
    with glucose 46
  in IDDM
    cholesterol levels 106, 117
    composition 46
    concentrations 46
    effect of insulin treatment 46
    kinetic studies of metabolism 46
  in NIDDM 47
    cholesterol levels 117
    composition 48
    effect of treatment 49
    kinetic studies of metabolism 48
  levels in diabetes mellitus 45

Watanabe heritable hyperlipidaemic rabbit 65